Letts EDUCATIONAL

ADVANCED LEVEL

Edmund Lewis

Revise A2 Chemistry

Author
Rob Ritchie

Contents

Specification lists

AQA Chemistry

MODULE	SPECIFICATION TOPIC	CHAPTER REFERENCE	STUDIED IN CLASS	REVISED	PRACTICE QUESTIONS
Module 4 (M4) *Further physical and organic chemistry*	*Kinetics*	*1.1*			
	Equilibria	*2.1, 2.2*			
	Acids and bases	*2.3–2.5*			
	Nomenclature and isomerism in organic chemistry	*5.1*			
	Compounds containing the carbonyl group	*5.2–5.5*			
	Aromatic chemistry	*6.1, 6.2*			
	Amines	*6.4*			
	Amino acids	*6.5*			
	Polymers	*6.6*			
	Organic synthesis and analysis	*7.4*			
	Structure determination	*7.7–7.3*			
Module 5 (M5) *Thermodynamics and further inorganic chemistry*	*Thermodynamics*	*3.1*			
	Periodicity	*4.1, 4.2*			
	Redox equilibria	*3.2, 3.3*			
	Transition metals	*4.3–4.7*			
	Reactions of inorganic compounds in aqueous solution	*4.8*			

Examination analysis

Units 1, 2 and 3 comprise AS Chemistry.			*50%*
A2 Chemistry comprises three unit tests. All questions are compulsory.			
Unit 4	*Structured questions: short and extended answers*	*1 hr 30 min test*	*15%*
Unit 5	*Structured questions: short and extended answers (includes synoptic assessment)*	*2 hr test*	*20%*
Unit 6 *comprises two components:*			
6(a)	*Objective questions: multiple-choice and either matching pairs* ***or*** *multiple completion items (synoptic assessment)*	*1 hr test*	*10%*
6(b)	*Centre-assessed coursework* ***or*** *practical examination*	*2 hr test*	*5%*

OCR Chemistry

MODULE	SPECIFICATION TOPIC	CHAPTER REFERENCE	STUDIED IN CLASS	REVISED	
Module 2814 (M4) Chains, rings and spectroscopy	*Arenes*	*6.1–6.3*			
	Carbonyl compounds	*5.2*			
	Carboxylic acids and esters	*5.3, 5.4*			
	Nitrogen compounds	*6.4, 6.5*			
	Stereoisomerism and organic synthesis	*5.1, 7.4*			
	Polymerisation	*6.6*			
	Spectroscopy	*7.1–7.3*			
Module 2815/1 (M5.1) Trends and patterns	*Lattice enthalpy*	*3.1*			
	The Periodic Table: Period 3	*4.1–4.2*			
	The Periodic Table: Transition elements	*4.3–4.8*			
Module 2815/2 (M5.2) Biochemistry					
Module 2815/2 (M5.3) Environmental chemistry					
Module 2815/4 (M5.4) Methods of analysis and detection		*7.1–7.3*			
Module 2815/5 (M5.5) Gases, liquids and solids					
Module 2815/6 (M5.6) Transition elements	*Electrode potentials*	*3.2, 3.3*			
	Ligands and complexes	*4.4*			
	Colour	*4.4*			
	Chemistry of Transition metals	*4.6, 4.7*			
Module 2816/1 (M6) Unifying concepts in chemistry	*How far?*	*1.1, 1.2*			
	How fast?	*2.1, 2.2*			
	Acids, bases and buffers	*2.3–2.5*			

Examination analysis

Units 2811, 2812 and 2813 comprise AS Chemistry			*50%*
A2 Chemistry comprises three unit tests. All questions are compulsory.			
Unit 2814	*Structured questions: short and extended answers*	*1 hr 30 min test*	*15%*
Unit 2815	*comprises two components:*		
2815/1	*Structured questions: short and extended answers (includes synoptic assessment)*	*1 hr test*	*7.5%*
2815/2–6	*One option only* Structured questions: short and extended answers*	*50 min test*	*7.5%*
Unit 2816	*comprises two components:*		
2816/1	*Structured questions: short and extended answers (synoptic assessment)*	*1 hr 15 min test*	*10%*
2816/2	*Centre-assessed coursework (includes synoptic assessment)* **or**		
2816/3	*Practical examination (includes synoptic assessment)*	*1 hr 30 min test*	*10%*

OCR Chemistry, Salters

MODULE	SPECIFICATION TOPIC	CHAPTER REFERENCE	STUDIED IN CLASS	REVISED	PRACTICE QUESTIONS
Module 2853 (M4) *Polymers, proteins and steel*	*Designer polymers*	*5.3–5.5, 6.4, 6.6*			
	Engineering proteins	*1.1, 2.1, 5.1, 6.5, 6.6, 7.3*			
	The steel story	*4.3–4.8, 3.2, 3.3*			
Module 2854 (M5) *Chemistry by design*	*Aspects of agriculture*	*1.1, 2.2, 3.2, 4.1, 4.2*			
	Colour by design	*5.4, 6.1, 6.2, 6.4*			
	The oceans	*2.3–2.5, 3.1*			
	Medicines by design	*5.2, 7.1–7.4*			

Note that the Salters approach 'drip-feeds' concepts throughout the course. You may need to revisit the chapters and practice questions in this Study Guide throughout your A2 course.

Examination analysis

Units 2850, 2851 and 2852 comprise AS Chemistry.			*50%*
A2 Chemistry comprises two unit tests. All questions are compulsory.			
Unit 2853	*Structured questions: short and extended answers*	*1 hr 30 min test*	*15%*
Unit 2854	*Structured questions: short and extended answers (includes synoptic assessment)*	*2 hr test*	*20%*
Unit 2855	*Individual investigation: internally assessed coursework (includes synoptic assessment)*		*15%*

Edexcel Chemistry

MODULE	SPECIFICATION TOPIC	CHAPTER REFERENCE	STUDIED IN CLASS	REVISED	PRACTICE QUESTIONS
Module 4 (M4) Periodicity, quantitative equilibria and functional group chemistry	Energetics II	3.1			
	Periodic Table II (Period 3 and Group 4)	4.1, 4.2			
	Chemical equilibria II	2.1, 2.2			
	Acid–base equilibria	2.3–2.5			
	Organic Chemistry II (acids, esters, carbonyl compounds, acid chlorides, nitrogen compounds and further halogeno compounds)	5.1–5.5, 6.4, 6.5			
Module 5 (M5) Transition metals, quantitative kinetics and applied organic chemistry	Redox equilibria (application)	3.2, 3.3			
	Transition metal chemistry	4.3–4.8			
	Organic chemistry III (reaction mechanisms and aromatic compounds)	6.1–6.4			
	Chemical kinetics II	1.1, 1.2			
	Organic chemistry IV (analysis, synthesis and applications)	7.1–7.4 6.6			

Examination analysis

Units 1, 2 and 3 comprise AS Chemistry		50%
A2 Chemistry comprises three unit tests.		
Unit 4	Structured questions: short answers. All questions are compulsory.	1 hr 30 min test 15%
Unit 5	Section A: structured questions Section B: structured and extended questions with a choice of questions. Parts of Section A and all of Section B will form part of the synoptic assessment.	1 hr 30 min test 15%
Unit 6 comprises two components:		
6A	Centre-assessed coursework or Practical examination	1 hr 45 min test 10%
6B	Synoptic paper Section A: Interpretation of data from laboratory situations Section B: Structured and extended questions with a choice of questions	1 hr 30 min test 10%

Edexcel Chemistry (Nuffield)

MODULE	SPECIFICATION TOPIC	CHAPTER REFERENCE	STUDIED IN CLASS	REVISED	PRACTICE QUESTIONS
Module 4 (M4) *Energy and reactions*	*How fast? Rates of reaction*	*1.1, 1.2*			
	Arenes: benzene and phenol	*6.1–6.3*			
	Entropy				
	How far? Reversible reactions	*2.1–2.5*			
	Oxidation products of alcohols	*5.2–5.5, 7.1*			
Module 5 (M5) *Special studies*	*Biochemistry*				
	Chemical engineering				
	Food science				
	Materials science				
	Mineral process chemistry				
Module 6 M6) *Applying chemistry*	*The Born–Haber cycle, structure and bonding*	*3.1, 4.1, 4.2*			
	Redox equilibria	*3.2, 3.3*			
	Natural and synthetic polymers	*5.1, 6.4–6.6*			
	The transition elements	*4.3–4.8*			
	Organic synthesis	*7.4*			
	Instrumental methods	*7.1–7.3*			

Examination analysis

Units 1, 2 and 3 comprise AS Chemistry.			*50%*
A2 Chemistry comprises three unit tests. All questions are compulsory.			
Unit 4	*Structured questions: short and extended answers*	*1 hr 30 min test*	*15%*
Unit 5	*Investigations and applications*		
Unit 5A	*Investigation, general practical competence*		*7.5%*
Unit 5B	*One option only* *Structured questions: short and extended answers*	*45 min test*	*7.5%*
Unit 6	*Synoptic (Open Book)*	*2 hr test*	*20%*

Unit 6 will contain structured questions, with a prerequisite of understanding of topics 1 to 15.

WJEC Chemistry

MODULE	SPECIFICATION TOPIC	CHAPTER REFERENCE	STUDIED IN CLASS	REVISED	PRACTICE QUESTIONS
Module CH4 (M4) *Spectroscopy and further organic chemistry*	*Spectroscopy*	*4.4, 7.1, 7.3*			
	Isomerism and aromaticity	*5.1, 6.1, 6.2*			
	Organic compounds containing halogens				
	Organic compounds containing oxygen	*5.2–5.5, 6.3*			
	Organic compounds containing nitrogen	*6.4, 6.5*			
	Organic synthesis and analysis	*6.6, 7.1–7.4*			
Module CH5 (M5) *Further physical and inorganic chemistry*	*Redox*	*3.2, 3.3, 4.6*			
	Chemistry of the s block				
	Chemistry of the p block				
	Transition elements	*4.3–4.8*			
	Periodicity	*4.1, 4.2*			
	Chemical kinetics	*1.1, 1.2*			
	Energy changes and equilibria	*2.1–2.5, 3.1*			

Examination analysis

Units 1, 2 and 3 comprise AS Chemistry. — *50%*

A2 Chemistry comprises three unit tests. All questions are compulsory.

Unit CH4	*Structured questions and objective questions*	*1 hr 40 min test*	*15%*
Unit CH5	*Structured questions and objective questions (includes synoptic assessment)*	*1 hr 40 min test*	*15%*

***Unit CH6** comprises two components:*

CH6a	*Structured questions: short and extended answers and comprehension (includes synoptic assessment)*	*1 hr 10 min test*	*10%*
CH6b	*Centre-assessed coursework (includes synoptic assessment)*		*10%*

NICCEA Chemistry

MODULE	SPECIFICATION TOPIC	CHAPTER REFERENCE	STUDIED IN CLASS	REVISED	PRACTICE QUESTIONS
Module 4 (M4) *Further organic, physical and inorganic chemistry*	*Lattice energy*	*3.1*			
	Kinetics	*1.1, 1.2*			
	Equilibrium	*2.1–2.5*			
	Electrode potentials	*3.2, 3.3*			
	Isomerism	*5.1*			
	Aldehydes and ketones	*5.2*			
	Carbohydrates				
	Carboxylic acids	*5.3, 5.5*			
	Esters, fats and oils	*5.4, 5.5, 6.6*			
	Periodic trends	*4.1, 4.2*			
	Oxy-acids of non-metals and their salts				
Module 5 (M5) *Analytical, transition metals and further organic chemistry*	*Analytical techniques*	*7.1–7.3*			
	Transition elements	*4.3–4.8*			
	Arenes	*6.1, 6.2*			
	Amines	*6.4*			
	Amino acids	*6.5, 6.6*			
	Nitriles	*7.4*			

Examination analysis

Units 1, 2 and 3 comprise AS Chemistry. *50%*

A2 Chemistry comprises three unit tests. All questions are compulsory.

Unit 4 *Objective questions and structured questions: short and extended answers (including synoptic assessment)* *1 hr 15 min test* *15%*

Unit 5 *Objective questions and structured questions: short and extended answers (including synoptic assessment)* *1 hr 15 min test* *15%*

***Unit 6** comprises two components:*

6A *Synoptic paper: structured questions: short and extended answers* *1hr 30 min test* *13.3%*

6B *Centre-assessed coursework* *6.7%*

AS/A2 Level Chemistry courses

AS and A2

All Chemistry A Level courses being studied from September 2000 are in two parts, with three separate modules in each part. Students first study the AS (Advanced Subsidiary) course. Some will then go on to study the second part of the A Level course, called A2. Advanced Subsidiary is assessed at the standard expected halfway through an A Level course: i.e., between GCSE and Advanced GCE. This means that new AS and A2 courses are designed so that difficulty steadily increases:

- AS Chemistry builds from GCSE Science
- A2 Chemistry builds from AS Chemistry.

How will you be tested?

Assessment units

For AS Chemistry, you will be tested by three assessment units. For the full A Level in Chemistry, you will take a further three units. AS Chemistry forms 50% of the assessment weighting for the full A Level.

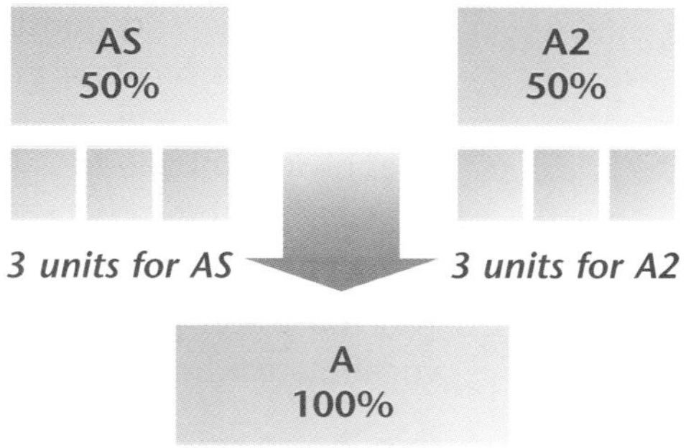

Each unit can normally be taken in either January or June. Alternatively, you can study the whole course before taking any of the unit tests.

If you are disappointed with a module result, you can resit each module once. You will need to be very careful about when you take up a resit opportunity because you will have only one chance to improve your mark. The higher mark counts.

Synoptic assessment

Synoptic assessment involves the explicit drawing together of knowledge, understanding and skills learned in different parts of the Advanced GCE course. The A2 units which draw together different parts of the course are assessed at the end of the course. More details are provided in Chapter 8 (pages 144–148).

Coursework

Coursework may form part of your A Level Chemistry course, depending on which specification you study. Where students have to undertake coursework, it is usually for the assessment of practical skills but this is not always the case. More details are provided on page 12.

Key skills

It is important that you develop your key skills of Communication, Application of Number and Information Technology throughout your AS and A2 courses. These are important skills that you need whatever you do beyond AS and A Levels. To gain the key skills qualification, you will need to demonstrate your ability to put your ideas across to other people, collect data and use up-to-date technology in your work. You will have many opportunities during A2 Chemistry to develop your key skills.

What skills will I need?

For A2 Chemistry, you will be tested by *assessment objectives*: these are the skills and abilities that you should have acquired by studying the course. The assessment objectives for A2 Chemistry are shown below.

Knowledge with understanding

- recall of facts, terminology and relationships
- understanding of principles and concepts
- drawing on existing knowledge to show understanding of the responsible use of chemistry in society
- selecting, organising and presenting information clearly and logically.

Application of knowledge and understanding, analysis and evaluation

- explaining and interpreting principles and concepts
- interpreting and translating, from one form into another, data presented as continuous prose or in tables, diagrams and graphs
- carrying out relevant calculations
- applying knowledge and understanding to familiar and unfamiliar situations
- assessing the validity of chemical information, experiments, inferences and statements.

You must also present arguments and ideas clearly and logically, using specialist vocabulary where appropriate.

Experimental and investigative skills

Chemistry is a practical subject and part of the assessment of A2 Chemistry will test your practical skills. This may be done during your lessons or you may be tested in a more formal practical examination. You will be assessed on four main skills:

- planning
- implementing
- analysing evidence and drawing conclusions
- evaluating evidence and procedures.

The skills may be assessed in separate practical exercises. They may also be assessed all together in the context of a single 'whole investigation'.

You will receive guidance about how your practical skills will be assessed from your teacher. This Study Guide concentrates on preparing you for the written examinations testing the subject content of A2 Chemistry.

AO4 Synthesis of knowledge, understanding and skills (Synoptic assessment)

- bringing together knowledge, principles and concepts from different areas of chemistry and apply them in a particular context
- using chemical skills in contexts which bring together different areas of the subject.

More details of synoptic assessment in Chemistry are discussed in Chapter 8 of this Study Guide (pages 144–148).

Different types of questions in A2 examinations

In A2 Chemistry examinations, different types of question are used to assess your abilities and skills. Unit tests mainly use structured questions requiring both short answers and more extended answers.

Short-answer questions

A short-answer question may test recall or it may test understanding. Short answer questions normally have space for the answers printed on the question paper.

Here are some examples (the answers are shown in *italics*):

What is meant by isotopes?

Atoms of the same element with different masses.

Calculate the amount (in mol) of H_2O in 4.5 g of H_2O.

1 mol H_2O has a mass of 18 g. ∴ 4.5 g of H_2O contains 4.5/18 = 0.25 mol H_2O.

Structured questions

Structured questions are in several parts. The parts usually have a common context and they often become progressively more difficult and more demanding as you work your way through the question. A structured question may start with simple recall, then test understanding of a familiar or an unfamiliar situation.

Most of the practice questions in this book are structured questions, as this is the main type of question used in the assessment of A2 Level Chemistry.

When answering structured questions, do not feel that you have to complete one question before starting the next. The further you are into a question, the more difficult the marks are to obtain. If you run out of ideas, go on to the next question. You need to respond to as many parts to questions on an exam paper as possible. You will not score well if you spend so long trying to perfect the first questions that you do not reach later questions at all.

Here is an example of a structured question that becomes progressively more demanding.

(a) Write down the atomic structure of the two isotopes of potassium: ^{39}K and ^{41}K.

(i) ^{39}K ...19... protons; ...20... neutrons; ...19... electrons. ✓

(ii) ^{41}K ...19... protons; ...22... neutrons; ...19... electrons. ✓ [2]

(b) A sample of potassium has the following percentage composition by mass: ^{39}K: 92%; ^{41}K: 8% Calculate the relative atomic mass of the potassium sample.

92 x 39/100 + 8 x 41/100 = 39.16 ✓ [1]

(c) What is the electronic configuration of a potassium atom?

$1s^2 2s^2 2p^6 3s^2 3p^6 4s^1$ ✓ [1]

(d) The second ionisation energy of potassium is much larger than its first ionisation energy.

(i) Explain what is meant by the *first ionisation energy* of potassium.

The energy required to remove an electron ✓ from each atom in 1 mole ✓ of gaseous atoms ✓

(ii) Why is there a large difference between the values for the first and the second ionisation energies of potassium?

The 2nd electron removed is from a different shell ✓ which is closer to the nucleus and experiences more attraction from the nucleus. ✓ This outermost electron experiences less shielding from the nucleus because there are fewer inner electron shells than for the 1st ionisation energy. ✓ [6]

Extended answers

In A2 Level Chemistry, questions requiring more extended answers may form part of structured questions or may form separate questions. They may appear anywhere on the paper and will typically have between 5 and 15 marks allocated to the answers as well as several lines of answer space. These questions are also often used to assess your abilities to communicate ideas and put together a logical argument.

The correct answers to extended questions are often less well-defined than to those requiring short answers. Examiners may have a list of points for which credit is awarded up to the maximum for the question.

An example of a question requiring an extended answer is shown below.

The table below gives data on the oxides of Period 3 in the Periodic Table.

oxide	Na_2O	MgO	Al_2O_3	SiO_2	P_4O_{10}	SO_3
melting point /K	1548	3100	2290	1880	853	306
conductivity of molten compound	good	good	medium	poor	poor	poor

Describe and explain the trends in formula, structure and bonding shown by these data. [11 marks]

Points that the examiners might look for include:

In the formula, the number of oxygen atoms bonded increases across the period. The oxidation number increases steadily by one for each successive element. ✓ This is because the number of electrons in the outer shell increases by one for each element in the period. ✓

The trend in structure is from giant to simple molecular ✓ with a changeover between silicon and phosphorus. ✓ Structure is related to the magnitude of the melting points. ✓

The high melting points (Na_2O ✓ SiO_2) result from strong forces between ions (for Na_2O – Al_2O_3) ✓ and between atoms (for SiO_2). ✓ The low melting points (P_4O_{10} ✓ SO_3) result from weak intermolecular forces or van der Waals' forces. ✓

The trend in bonding is from ionic to covalent ✓ with a changeover between Al_2O_3 and SiO_2. ✓ This is related to the conductivity of the molten oxide. ✓

Good conductivity results from mobile ions (Na_2O ✓ $Al2O_3$. ✓)

Poor conductivity shows that no mobile ions are present (SiO_2 ✓ SO_3. ✓)

13 marking points → 11 marks

In this type of response, there may be an additional mark for a clear, well-organised answer, using specialist terms. In addition, marks may be allocated for legible text with accurate spelling, punctuation and grammar.

Other types of questions

Free-response and open-ended questions allow you to choose the context and to develop your own ideas.

Multiple-choice or objective questions require you to select the correct response to the question from a number of given alternatives.

Exam technique

Links from AS

A2 Chemistry builds from the knowledge and understanding acquired after studying AS Chemistry. This Study Guide has been written so that you will be able to tackle A2 Chemistry from an AS Chemistry background.

You should not need to search for important chemistry from AS Chemistry because cross-references have been included as 'Key points from AS' to the AS Chemistry Study Guide: Revise AS.

What are examiners looking for?

Examiners use instructions to help you to decide the length and depth of your answer.

If a question does not seem to make sense, you may have misread it – read it again!

State, define or list

This requires a short, concise answer, often recall of material that can be learnt by rote.

Explain, describe or discuss

Some reasoning or some reference to theory is required, depending on the context.

Outline

This implies a short response, almost a list of sentences or bullet points.

Predict or deduce

You are not expected to answer by recall but by making a connection between pieces of information.

Suggest

You are expected to apply your general knowledge to a 'novel' situation, one which you have not directly studied during the AS Chemistry course.

Calculate

This is used when a numerical answer is required. You should always use units in quantities and significant figures should be used with care.

Look to see how many significant figures have been used for quantities in the question and give your answer to this degree of accuracy.

If the question uses 3 sig figs, then give your answer to 3 sig figs also.

Some dos and don'ts

Dos

Do answer the question.

- No credit can be given for good Chemistry that is irrelevant to the question.

Do use the mark allocation to guide how much you write.

- Two marks are awarded for two valid points - writing more will rarely gain more credit and could mean wasted time or even contradicting earlier valid points.

Do use diagrams, equations and tables in your responses.

- Even in 'essay-type' questions, these offer an excellent way of communicating chemistry.

Do write legibly.

- An examiner cannot give marks if the answer cannot be read.

Do write using correct spelling and grammar. Structure longer essays carefully.

- Marks are now awarded for the quality of your language in exams.

Don'ts

Don't fill up any blank space on a paper.

- In structured questions, the number of dotted lines should guide the length of your answer.
- If you write too much, you waste time and may not finish the exam paper. You also risk contradicting yourself.

Don't write out the question again.

- This wastes time. The marks are for the answer!

Don't contradict yourself.

- The examiner cannot be expected to choose which answer is intended.

Don't spend too much time on a part that you find difficult.

- You may not have enough time to complete the exam. You can always return to a difficult calculation if you have time at the end of the exam.

What grade do you want?

Everyone would like to improve their grades but you will only manage this with a lot of hard work and determination. You should have a fair idea of your natural ability and likely grade in chemistry and the hints below offer advice on improving that grade.

For a Grade A

You will need to be a very good all-rounder.

- You must go into every exam knowing the work extremely well.
- You must be able to apply your knowledge to new, unfamiliar situations.
- You need to have practised many, many exam questions so that you are ready for the type of question that will appear.

The exams test all areas of the specification and any weaknesses in your chemistry will be found out. There must be no holes in your knowledge and understanding. For a Grade A, you must be competent in all areas.

For a Grade C

You must have a reasonable grasp of chemistry but you may have weaknesses in several areas and you will be unsure of some of the reasons for the chemistry.

- Many Grade C candidates are just as good at answering questions as Grade A students but holes and weaknesses often show up in just some topics.
- To improve, you will need to master your weaknesses and you must prepare thoroughly for the exam. You must become a better all-rounder.

For a Grade E

You cannot afford to miss the easy marks. Even if you find chemistry difficult to understand and would be happy with a Grade E, there are plenty of questions in which you can gain marks.

- You must memorise all definitions.
- You must practise exam questions to give yourself confidence that you do know some chemistry. In exams, answer the parts of questions that you know first. You must not waste time on the difficult parts. You can always go back to these later.
- The areas of chemistry that you find most difficult are going to be hard to score on in exams. Even in the difficult questions, there are still marks to be gained. Show your working in calculations because credit is given for a sound method. You can always gain some marks if you get part of the way towards the solution.

What marks do you need?

As a rough guide, you will need to score an average of 40% for a Grade E, 60% for a Grade C and 80% for a Grade A.

average	*80%*	*70%*	*60%*	*50%*	*40%*
grade	*A*	*B*	*C*	*D*	*E*

- When you get your results, identify a realistic 'target grade' for the A Level from your average mark.

Four steps to successful revision

Step 1: Understand

- Study the topic to be learned slowly. Make sure you understand the logic or important concepts.
- Mark up the text if necessary – underline, highlight and make notes.
- Re-read each paragraph slowly.

GO TO STEP 2

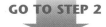

Step 2: Summarise

- Now make your own revision note summary:
  What is the main idea, theme or concept to be learned?
  What are the main points? How does the logic develop?
  Ask questions: Why? How? What next?
- Use bullet points, mind maps, patterned notes.
- Link ideas with mnemonics, mind maps, crazy stories.
- Note the title and date of the revision notes
  (e.g. Chemistry: Reaction rates, 3rd March).
- Organise your notes carefully and keep them in a file.

This is now in **short-term memory**. You will forget 80% of it if you do not go to Step 3.
GO TO STEP 3, but first take a 10 minute break.

Step 3: Memorise

- Take 25 minute learning 'bites' with 5 minute breaks.
- After each 5 minute break test yourself:
  Cover the original revision note summary
  Write down the main points
  Speak out loud (record on tape)
  Tell someone else
  Repeat many times.

The material is well on its way to **long-term memory**.
You will forget 40% if you do not do step 4. **GO TO STEP 4**

Step 4: Track/Review

- Create a Revision Diary (one A4 page per day).
- Make a revision plan for the topic, e.g. 1 day later, 1 week later, 1 month later.
- Record your revision in your Revision Diary, e.g.
  Chemistry: Reaction rates, 3rd March 25 minutes
  Chemistry: Reaction rates, 5th March 15 minutes
  Chemistry: Reaction rates, 3rd April 15 minutes
  ... and then at monthly intervals.

Chapter 1

Reaction rates

The following topics are covered in this chapter:

- *Orders and the rate equation*
- *Determination of reaction mechanisms*

1.1 Orders and the rate equation

After studying this section you should be able to:

- *explain and use the term 'rate of reaction'*
- *explain and use the terms 'order', 'rate constant', 'rate equation'*
- *understand that a temperature change increases the rate constant*
- *use the initial rates method to deduce orders*
- *construct a rate equation and calculate its rate constant*
- *use graphs to deduce reaction rates and orders*

LEARNING SUMMARY

Key points from AS

- **Reaction rates**
  Revise AS pages 98–101

During the study of reaction rates in AS Chemistry, reaction rates were described in terms of activation energy and the Boltzmann distribution. For A2 Chemistry, you will build upon this knowledge and understanding by measuring and calculating reaction rates using rate equations.

Orders and the rate equation

AQA	M4	SALTERS	M4, M5
EDEXCEL	M5	WJEC	CH5
OCR	M6	NICCEA	M4
NUFFIELD	M4		

The rate of a reaction is usually measured as the **change in concentration** of a reaction species **with time**.

- Units of rate = $\underbrace{\text{mol dm}^{-3}}_{\text{concentration}}\ \underbrace{\text{s}^{-1}}_{\text{per time}}$

concentration   per time

Key points from AS

- **What is a reaction rate?**
  Revise AS pages 98–99

[**A**] means the concentration of reactant **A** in mol dm^{-3}.

Orders of reaction

For a reaction: **A** + **B** + **C** ⟶ products,

- the **order** of the reaction shows how the reaction rate is affected by the concentrations of **A**, **B** and **C**.

If the order is 0 (zero order) with respect to reactant **A**,

- the rate is unaffected by changes in concentration of **A**:
  $rate \propto [\mathbf{A}]^0$

If the order is 1 (first order) with respect to a reactant **B**,

- the rate is doubled by doubling of the concentration of reactant **B**:
  $rate \propto [\mathbf{B}]^1$

If the order is 2 (second order) with respect to a reactant **C**,

- the rate is quadrupled by doubling of the concentration of reactant **C**:
  $rate \propto [\mathbf{C}]^2$

Combining the information above,

$rate \propto [\mathbf{A}]^0[\mathbf{B}]^1[\mathbf{C}]^2$

$\therefore\ rate = k[\mathbf{B}]^1[\mathbf{C}]^2$

- This expression is called the **rate equation** for the reaction.
- k is the rate constant of the reaction.

A number raised to the power of zero = 1 and zero order species can be omitted from the rate equation.

The rate equation

The rate of a reaction is determined from experimental results.

It **cannot** be determined from the chemical equation.

The rate equation of a reaction shows how the rate is affected by the concentration of reactants. A rate equation can only be determined from experiments.

In general, for a reaction: **A** + **B** ⟶ **C**,
the reaction rate is given by: *rate* = $k[\mathbf{A}]^m[\mathbf{B}]^n$

- m and n are the **orders of reaction** with respect to **A** and **B** respectively.
- The **overall order** of reaction is m + n.
- The reaction rate is measured as the change in concentration of a reaction species with time.
- The units of rate are mol dm^{-3} s^{-1}.

The rate constant *k* indicates the rate of the reaction:
a large value of *k* ⟶ fast rate of reaction
a small value of *k* ⟶ slow rate of reaction.

The effect of temperature on rate constants

Key points from AS

- **Activation energy** *Revise AS page 99*

An increase in temperature speeds up the rate of most reactions by **increasing** the rate constant *k*.

The table below shows the increase in the rate constant for the decomposition of hydrogen iodide with increasing temperature.

$$2HI(g) \longrightarrow H_2(g) + I_2(g)$$

temperature/°C	283	356	427	508
rate constant, *k* / dm^3 mol^{-1} s^{-1}	7.04×10^{-7}	6.04×10^{-5}	2.32×10^{-3}	7.90×10^{-2}

For many reactions, the rate doubles for each 10°C increase in temperature.

A reaction with a large activation energy has a **small rate constant**. Such a reaction may need a considerable temperature rise to increase the value of the rate constant sufficiently for a reaction to take place at all.

Measuring rates using the initial rates method

AQA	M4	SALTERS	M4
EDEXCEL	M5	WJEC	CH5
OCR	M6	NICCEA	M4
NUFFIELD	M4		

For a reaction, **X** + **Y** ⟶ **Z**,

a series of experiments can be carried out using different **initial** concentrations of the reactants **X** and **Y**.

It is important to change only one variable at a time, so two series of experiments will be required.

- **Series 1** – The concentration of **X is changed** whilst the concentration of **Y is kept constant.**
- **Series 2** – The concentration of **Y is changed** whilst the concentration of **X is kept constant.**

For each experiment, we need to:

- plot a **concentration/time** graph
- measure the **initial rate** from the graph as the **tangent** drawn at **time = 0** (see page 23).

The initial rate of the reaction is the tangent of a concentration/time graph at time = 0.

The results below show the initial rates using different concentrations of **X** and **Y**.

Experiment	[X(aq)] /mol dm^{-3}	[Y(aq)] /mol dm^{-3}	initial rate /mol dm^{-3} s^{-1}
1	1.0×10^{-2}	1.0×10^{-2}	0.5×10^{-3}
2	2.0×10^{-2}	1.0×10^{-2}	2.0×10^{-3}
3	2.0×10^{-2}	2.0×10^{-2}	4.0×10^{-3}

Determine the orders

Comparing experiments 1 and 2: [**Y**(aq)] has been kept constant

- [**X**(aq)] has been doubled, rate × 4
  ∴ order with respect to **X**(aq) = 2.

Comparing experiments 2 and 3: [**X**(aq)] has been kept constant

- [**Y**(aq)] has been doubled, rate doubles
  ∴ order with respect to **Y**(aq) = 1.

Use the orders to write the Rate Equation

- $rate = k[\mathbf{X}(aq)]^2[\mathbf{Y}(aq)]$
- The overall order of this reaction is (2 + 1) = 3rd order

Calculate the rate constant for the reaction

Rearrange the rate equation:

$$\text{The rate constant, } k = \frac{rate}{[\mathbf{X}(aq)]^2[\mathbf{Y}(aq)]}$$

Calculate k using values from one of the experiments.

$$k = \frac{(2.0 \times 10^{-3})}{(2.0 \times 10^{-2})^2\,(1.0 \times 10^{-2})} = 500 \text{ dm}^6 \text{ mol}^{-2} \text{ s}^{-1}$$

In this example, the results from Experiment 2 have been used.

You will get the same value of k from the results of **any** of the experimental runs.

Try calculating k from Experiment 1 and from Experiment 3. All give the same value.

Units of rate constants

The units of a rate constant depend upon the rate equation for the reaction. We can determine units of k by substituting units for rate and concentration into the rearranged rate equation. The table below shows how units can be determined.

For a **zero order** reaction: $rate = k[\mathbf{A}]^0 = k$ units of k = $\mathbf{mol\ dm^{-3}\ s^{-1}}$	For a **first order** reaction: $rate = k[\mathbf{A}]$ ∴ $k = \frac{rate}{[\mathbf{A}]}$ Units of $k = \frac{(\text{mol dm}^{-3}\text{ s}^{-1})}{(\text{mol dm}^{-3})} = \mathbf{s^{-1}}$
For a **second order** reaction: $rate = k[\mathbf{A}]^2$ ∴ $k = \frac{rate}{[\mathbf{A}]^2}$ Units of $k = \frac{(\text{mol dm}^{-3}\text{ s}^{-1})}{(\text{mol dm}^{-3})^2}$ $= \mathbf{dm^3\ mol^{-1}\ s^{-1}}$	For a **third order** reaction: $rate = k[\mathbf{A}]^2[\mathbf{B}]$ ∴ $k = \frac{rate}{[\mathbf{A}]^2[\mathbf{B}]}$ Units of $k = \frac{(\text{mol dm}^{-3}\text{ s}^{-1})}{(\text{mol dm}^{-3})^2\,(\text{mol dm}^{-3})}$ $= \mathbf{dm^6\ mol^{-2}\ s^{-1}}$

Notice how units need to be worked out afresh for reactions with different overall orders.

You do not need to memorise these but it is important that you are able to work these out when needed.

See also p. 30 in which the units of the equilibrium constant K_c are discussed.

Progress check

1 A chemical reaction is first order with respect to compound **P** and second order with respect to compound **Q**.

(a) Write the rate equation for this reaction.

(b) What is the overall order of this reaction?

(c) By what factor will the rate increase if:

(i) the concentration of **P only** is doubled

(ii) the concentration of **Q only** is doubled

(iii) the concentrations of **P** and **Q** are **both** doubled?

(d) What are the units of the rate constant of this reaction?

2 The reaction of ethanoic anhydride, $(CH_3CO)_2O$, with ethanol, C_2H_5OH, can be represented by the equation:

$$(CH_3CO)_2O + C_2H_5OH \longrightarrow CH_3CO_2C_2H_5 + CH_3CO_2H$$

The table below shows the initial concentrations of the two reactants and the initial rates of reaction.

Experiment	*$[(CH_3CO)_2O]$ /mol dm^{-3}*	*$[C_2H_5OH]$ /mol dm^{-3}*	*initial rate /mol dm^{-3} s^{-1}*
1	0.400	0.200	6.60×10^{-4}
2	0.400	0.400	1.32×10^{-3}
3	0.800	0.400	2.64×10^{-3}

(a) State and explain the order of reaction with respect to:
(i) ethanoic anhydride (ii) ethanol.

(b) (i) Write an expression for the overall rate equation.
(ii) What is the overall order of reaction?

(c) Calculate the value, with units, for the rate constant, *k*.

1 (a) *rate* = $k[P][Q]^2$
(b) 3rd order
(c) (i) rate doubles (ii) rate quadruples (iii) rate × 8
(d) dm^6 mol^{-2} s^{-1}

2 (a) (i) 1st order; concentration doubles; rate doubles
(ii) 1st order; concentration doubles; rate doubles.
(b) (i) *rate* = $k\,[(CH_3CO)_2O]\,[C_2H_5OH]$ (ii) 2nd order.
(c) 8.25×10^{-3} dm^3 mol^{-1} s^{-1}.

Measuring rates using graphs

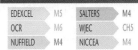

EDEXCEL	M5	SALTERS	M4
OCR	M6	WJEC	CH5
NUFFIELD	M4	NICCEA	M4

Concentration/time graphs

It is often possible to measure the concentration of a reactant or product at various times during the course of an experiment.

- From the results, a concentration/time graph can be plotted.
- The shape of this graph can indicate the order of the reaction by measuring the **half-life** of a reactant.

Key points from AS

- **What is a reaction rate?**
Revise AS pages 98–99

KEY POINT

The half-life of a reactant is the time taken for its concentration to reduce by half.

- **A first-order reaction has a constant half-life.**

Example of a first order graph

The reaction between bromine and methanoic acid is shown below.

$Br_2(aq) + HCOOH(aq) \longrightarrow 2Br^-(aq) + 2H^+(aq) + CO_2(g)$

- During the course of the reaction, the orange bromine colour disappears as Br_2 reacts to form colourless Br^- ions. The order with respect to bromine can be determined by monitoring the rate of disappearance of bromine using a colorimeter.
- The concentration of the other reactant, methanoic acid, is kept virtually constant by using an excess of methanoic acid.

The concentration/time graph from this reaction is shown on the next page. Notice that the **half-life is constant** at 200 s, showing that this reaction is first order with respect to $Br_2(aq)$.

The half-life curve of a 1st order reaction is concentration independent.

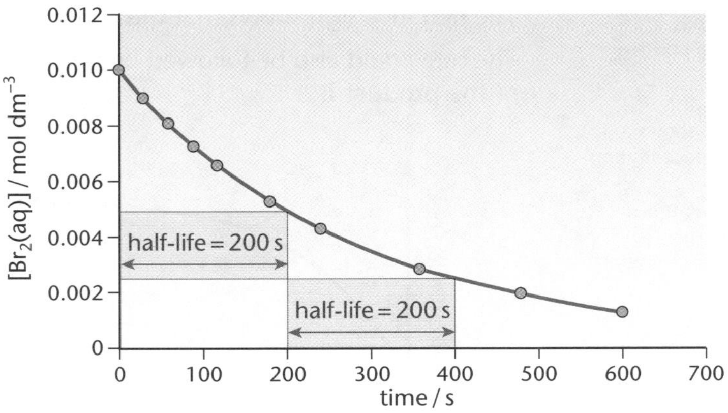
0.012
0.010
0.008
0.006
0.004
0.002
0
[Br$_2$(aq)] / mol dm^{-3}
half-life = 200 s
half-life = 200 s
0
100
200
300
400
500
600
700
time / s

The shapes of concentration/time graphs for zero, first and second order reactions are shown below.

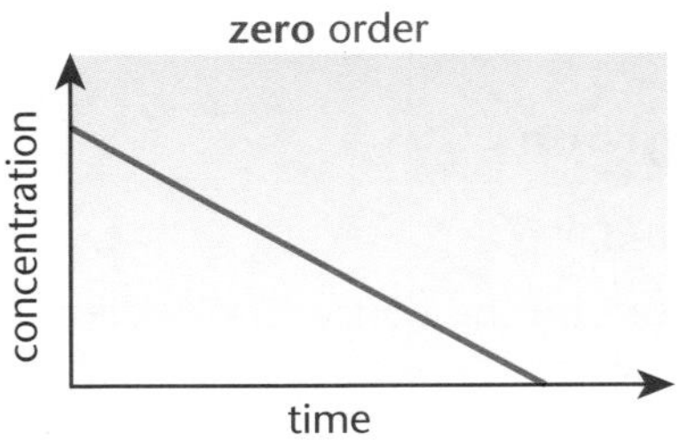
zero order
concentration
time

The concentration falls at a steady rate with time

- half-life **decreases** with time.

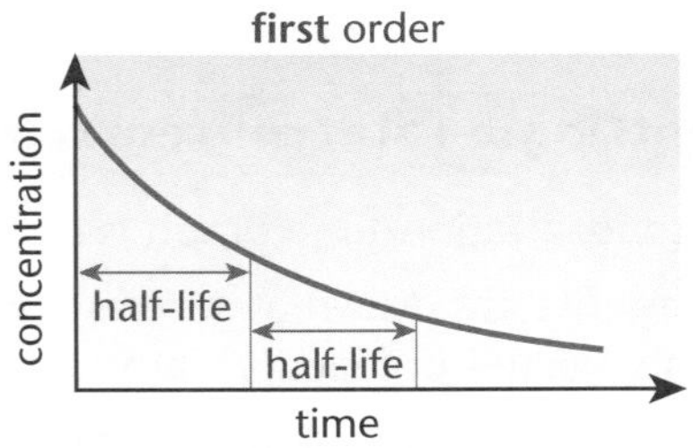
first order
concentration
half-life
half-life
time

The concentration halves in equal time intervals

- **constant** half-life.

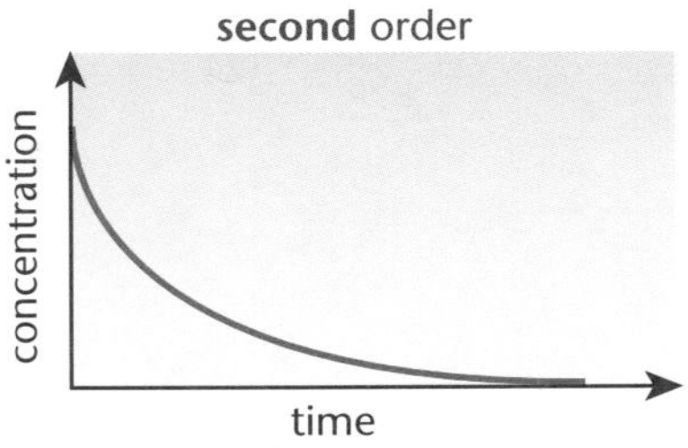
second order
concentration
time

The half-life gets progressively longer as the reaction proceeds

- half-life **increases** with time.

- The 1st order relationship can be confirmed by plotting a graph of log[**X**] against time, which gives a straight line.

Measuring rates from tangents

The gradient of the concentration/time graph is a measure of the rate of a reaction.

For the reaction: **A** ⟶ **B**, the graph below shows how the concentration of **A** changes during the course of the reaction.

KEY POINT

At any instant of time during the reaction, the rate can be measured by drawing a tangent to the curve.

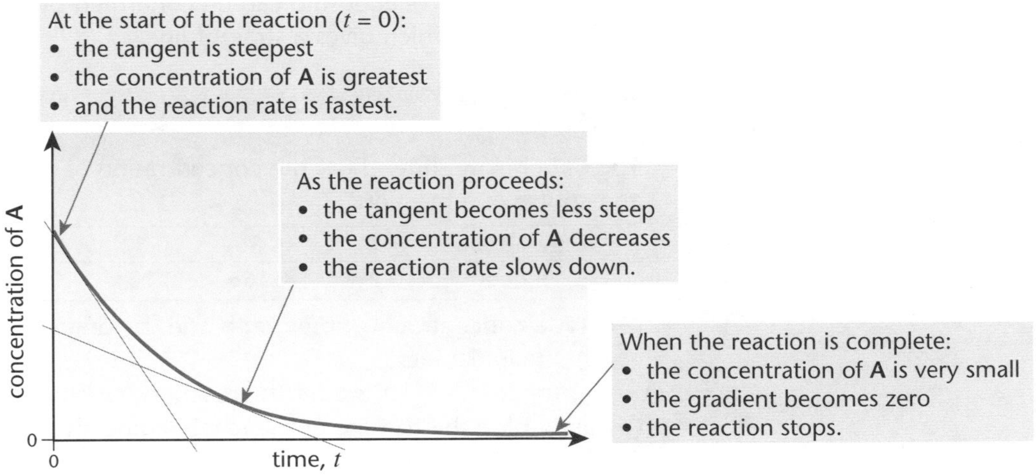
At the start of the reaction (t = 0):
• the tangent is steepest
• the concentration of A is greatest
• and the reaction rate is fastest.
As the reaction proceeds:
• the tangent becomes less steep
• the concentration of A decreases
• the reaction rate slows down.
When the reaction is complete:
• the concentration of A is very small
• the gradient becomes zero
• the reaction stops.
concentration of A
0
0
time, t

At an instant of time, t, during the reaction:

- the rate of **decrease** in concentration of **A** = $-\frac{d[A]}{dt}$ *(the tangent)*

A reaction rate is expressed as a positive value.

For the reactant **A**, the **negative** sign shows a **decreasing** concentration with time.

For the product **B**, the **positive** sign shows **increasing** concentration.

The negative sign shows that the concentration of **A** decreases during the reaction.

The rate could also be followed by measuring the rate of **increase** in concentration of the product **B**.

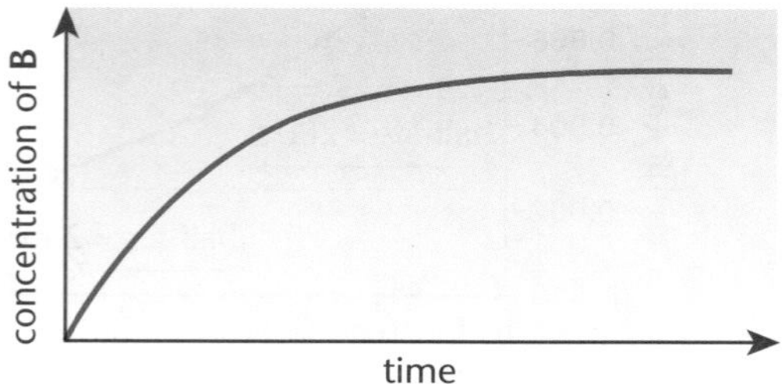

B is formed and its concentration increases during the reaction.

$$\text{rate of } \textbf{increase} \text{ in concentration of } \textbf{B} = +\frac{d[\textbf{B}]}{dt}$$

Plotting a rate/concentration graph

- A concentration/time graph is first plotted.
- **Tangents** are drawn at several time values on the concentration/time graph, giving values of reaction rates.
- A second graph is plotted of **rate** against **concentration**.
- The shape of this graph confirms the order of the reaction.

zero order

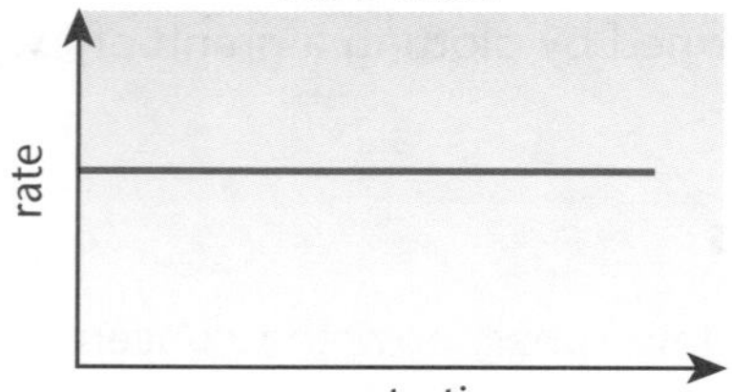

Rate $\propto$ [A]^0 $\therefore$ rate = constant
- Rate unaffected by changes in concentration.

first order

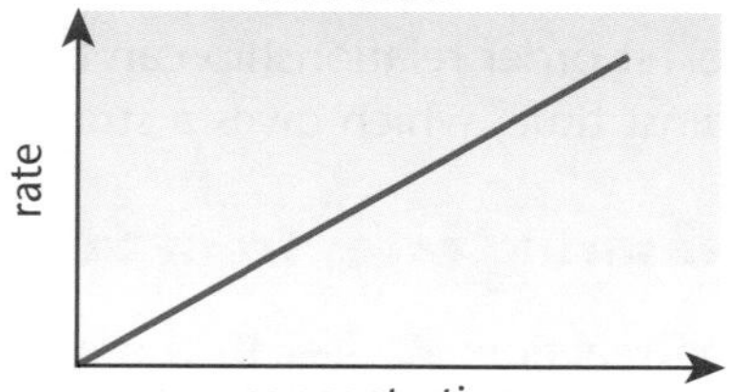

Rate $\propto$ [A]^1
- Rate doubles as concentration doubles.

second order

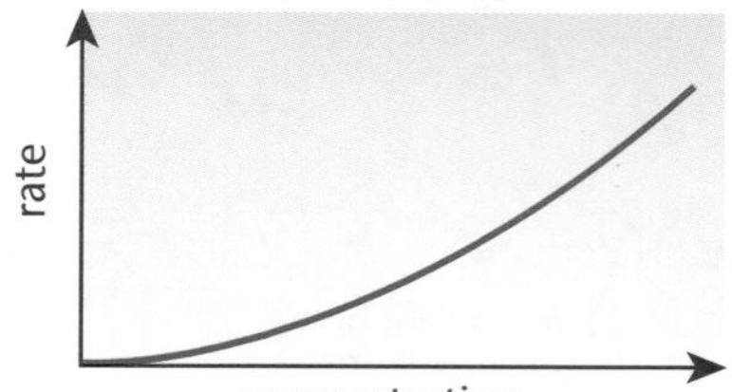

Rate $\propto$ [A]^2
- Rate quadruples as concentration doubles.

- The 2nd order relationship can be confirmed by plotting a graph of rate against [A]^2, which gives a straight line.

Progress check

1 The data below shows how the concentration of a reactant **A** changes during the course of a reaction.

Time / 10^4 s	0	0.36	0.72	1.08	1.44
[A] /mol dm^{-3}	0.240	0.156	0.104	0.068	0.045

(a) Plot a concentration/time graph and, by drawing tangents, estimate:
  (i) the initial rate
  (ii) the rate 1 × 10^4 s after the reaction has started.

(b) Determine the half-life of the reaction and show that the reaction is first order with respect to **A**.

1 (a) (i) 2.9 × 10^{-5} mol dm^{-3} s^{-1} (ii) 8.7 × 10^{-6} mol dm^{-3} s^{-1}
(b) Half-life = 5.8 × 10^3 s. Successive half-lives are constant.

1.2 Determination of reaction mechanisms

After studying this section you should be able to:

- *understand what is meant by a rate-determining step*
- *predict a rate equation from a rate-determining step*
- *predict a rate-determining step from a rate equation*
- *use a rate equation and the overall equation for a reaction to predict a possible reaction mechanism*

LEARNING SUMMARY

Predicting reaction mechanisms

EDEXCEL	M5	WJEC	CH5
OCR	M6	NICCEA	M4
NUFFIELD	M4		

The rate-determining step

Chemical reactions often take place in a series of steps. The detail of these steps is the **reaction mechanism**.

> The rate equation can provide clues about a likely reaction mechanism by identifying the **slowest** stage of a reaction sequence, called the **rate-determining step**.
>
> KEY POINT

Key points from AS

- **The hydrolysis of halogenoalkanes**
  Revise AS page 132–133

Predicting reaction mechanisms from rate equations

Halogenoalkanes are hydrolysed by hot aqueous alkali.

$$RBr + OH^- \longrightarrow ROH + Br^-$$

Rates for the hydrolysis reactions of primary and tertiary halogenoalkanes can be measured experimentally to determine rate equations. We can predict the rate-determining step for each of these reactions from the rate equations obtained.

The hydrolysis of primary halogenoalkanes

A balanced equation represents the **overall** reaction.

It does not reveal the mechanism that achieves it.

The primary halogenoalkane, $CH_3CH_2CH_2CH_2Br$, is hydrolysed by aqueous alkali:

$$CH_3CH_2CH_2CH_2Br + OH^- \longrightarrow CH_3CH_2CH_2CH_2OH + Br^-$$

Experiments show that the rate equation for this reaction is:

$$\textit{rate} = k[CH_3CH_2CH_2CH_2Br]\,[OH^-]$$

This rate equation shows that:

- the rate is determined by a **slow reaction step** which must involve **both $CH_3CH_2CH_2CH_2Br$ and OH^-**.

This supports the **single step** mechanism below, matching the rate equation.

$$CH_3CH_2CH_2CH_2Br + OH^- \longrightarrow CH_3CH_2CH_2CH_2OH + Br^-$$

$$\downarrow \qquad \downarrow$$

$$\textit{rate} = k[CH_3CH_2CH_2CH_2Br]\,[OH^-]$$

Evidence for the mechanism comes from the rate equation.

The hydrolysis of tertiary halogenoalkanes

The tertiary halogenoalkane, $(CH_3)_3CBr$, is hydrolysed by aqueous alkali:

$$(CH_3)_3CBr + OH^- \longrightarrow (CH_3)_3COH + Br^-$$

Experiments show that the rate equation for this reaction is:

$$\textit{rate} = k[(CH_3)_3CBr]$$

This rate equation shows that:

- the rate is determined by a slow reaction step which involves **only $(CH_3)_3CBr$**
- the concentration of OH^- has **no effect** on the reaction rate.

A possible step in the reaction mechanism including only $(CH_3)_3CBr$ is:

$$(CH_3)_3CBr \longrightarrow (CH_3)_3C^+ + Br^-$$
$$\downarrow$$
$$rate = k[(CH_3)_3CBr]$$

This supports the **two step** mechanism below, matching the rate equation.

Step 1: the rate-determining step

$$(CH_3)_3CBr \longrightarrow (CH_3)_3C^+ + Br^- \qquad \textbf{SLOW}$$

Step 2

$$(CH_3)_3C^+ + OH^- \longrightarrow (CH_3)_3COH \qquad \textbf{FAST}$$

In this reaction sequence, the rate is determined by the **slow first step**.

The rate equation only includes the reacting species for this rate-determining step.

$$rate = k[(CH_3)_3CBr]$$

- The rate of the reaction is controlled mainly by the slowest step of the mechanism – the **rate-determining step**.
- The rate equation only includes reacting species involved in the slow rate-determining step.
- The orders in the rate equation match the number of species involved in the rate-determining step.

You are **not** expected to remember these examples.

You **are** expected to interpret data to identify a rate-determining step and to suggest a possible reaction mechanism for a reaction.

Progress check

1 CH_3CH_2Br and cyanide ions, CN^- react as follows:

$$CH_3CH_2Br + CN^- \longrightarrow CH_3CH_2CN + Br^-$$

Show two different possible reaction routes for this reaction and write down the expected rate equation for each route.

1 Single step: $CH_3CH_2Br + CN^- \longrightarrow CH_3CH_2CN + Br^-$

rate = $k[CH_3CH_2Br][CN^-]$

One step: $CH_3CH_2Br + CN^- \longrightarrow CH_3CH_2CN + Br^-$

Two steps: $CH_3CH_2Br \longrightarrow CH_3CH_2^+ + Br^-$ slow

$CH_3CH_2^+ + CN^- \longrightarrow CH_3CH_2CN$ fast

Sample question and model answer

Note the key points for the marks:
1 mark for line going through the origin
1 mark for the graph levelling off
1 mark for showing how the initial rate is measured.

The oxidation of nitrogen monoxide to nitrogen dioxide in car exhausts may involve carbon monoxide and oxygen.

$$NO(g) + CO(g) + O_2(g) \longrightarrow NO_2(g) + CO_2(g)$$

The rate of this reaction can be followed colorimetrically because $NO_2(g)$ is coloured.

(a) (i) Use the axes below to show how the concentration of $NO_2(g)$ produced varies with time.

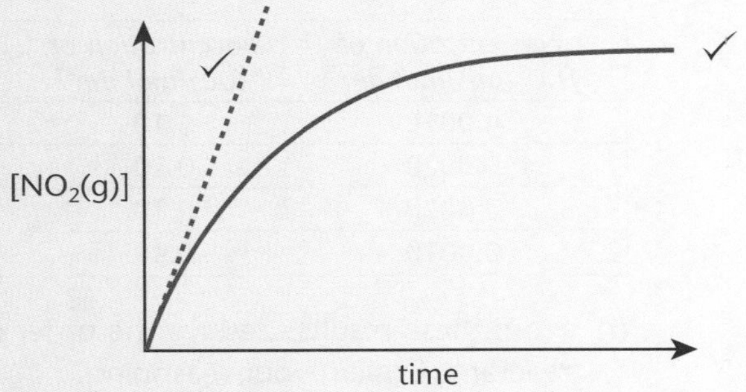

(ii) Show, with reference to your sketch, how the initial rate of reaction could be deduced.

Draw a tangent to the curve at time = 0 s (see above in diagram). ✓ [3]

(b) The results from a series of experiments are shown below:

Experiment	*[NO(g)] /mol dm^{-3}*	*[CO(g)] /mol dm^{-3}*	*[O$_2$(g)] /mol dm^{-3}*	*Initial rate /mol dm^{-3} s^{-1}*
1	1.00×10^{-3}	1.00×10^{-3}	1.00×10^{-1}	4.40×10^{-4}
2	2.00×10^{-3}	1.00×10^{-3}	1.00×10^{-1}	1.76×10^{-3}
3	2.00×10^{-3}	2.00×10^{-3}	1.00×10^{-1}	1.76×10^{-3}
4	4.00×10^{-3}	1.00×10^{-3}	2.00×10^{-1}	7.04×10^{-3}

Straightforward marks. Show your logic.

(i) Deduce the order of reaction with respect to NO(g). Show your reasoning.

double concentration of NO only, rate quadruples. ✓
∴ second order with respect to NO. ✓

(ii) State the order of reaction with respect to CO(g) and to O_2(g).

CO: double concentration of CO only, rate does not change:
∴ zero order with respect to CO. ✓
O_2: double concentration of NO AND O_2, rate quadruples.
The effect of doubling [NO] quadruples the rate.
∴ changing [O_2] has no effect and zero order with respect to O_2. ✓

(iii) Write an expression for the rate equation.

rate = $k[NO(g)]^2$ ✓

You can use any of the results – they all give the same value for *k*.

Substitute units into the rate expression to find the units of *k*.

(iv) Calculate a value for the rate constant, *k*. State the units of *k*.

Using results from the 1st experiment: $4.40 \times 10^{-4} = k \times (1.00 \times 10^{-3})^2$ ✓

$$\therefore k = \frac{4.40 \times 10^{-4}}{(1.00 \times 10^{-3})^2} = 440$$ ✓

units: $\frac{mol\ dm^{-3}\ s^{-1}}{(mol\ dm^{-3})^2} = mol^{-1}\ dm^3\ s^{-1}$ ✓ $\quad k = 440\ dm^3\ mol^{-1}\ s^{-1}$ [8]

[Total: 11]

Cambridge How Far, How Fast? Q1 June 1996

Practice examination questions

1

The reaction between hydrogen peroxide and iodide ions in an acid solution can be written as follows.

$$H_2O_2(aq) + 2I^-(aq) + 2H^+(aq) \longrightarrow I_2(aq) + 2H_2O(l)$$

A student wanted to investigate the rate of this reaction.

(a) Suggest how the student could follow the rate of this reaction in the laboratory. [2]

(b) The student obtained the following results.

concentration of $H_2O_2(aq)$/mol dm^{-3}	*concentration of $I^-(aq)$/mol dm^{-3}*	*concentration of $H^+(aq)$/mol dm^{-3}*	*initial rate /10^{-6} mol dm^{-3} s^{-1}*
0.0010	0.10	0.10	2.8
0.0020	0.10	0.10	5.6
0.0020	0.10	0.20	5.6
0.0010	0.40	0.10	11.2

(i) From these results, deduce the order of reaction with respect to each reactant. Explain your reasoning. [6]

(ii) Using your answers to (b) (i), write a rate equation for this reaction. [1]

(iii) Calculate the value of the rate constant, *k*, for this reaction. [2]

(c) (i) What is meant by the term rate-determining step in a reaction sequence? [1]

(ii) Using your answer to (b) (ii), suggest an equation for the rate-determining step for the reaction between H_2O_2 and I^- in the presence of acid. [2]

[Total: 14]

Cambridge How Far, How Fast? Q2 March 2000

2

(a) A chemical reaction is first order with respect to compound **X** and second order with respect to compound **Y**.

(i) Write the rate equation for this reaction. [2]

(ii) What is the overall order of this reaction? [1]

(iii) By what factor will the rate increase if the concentrations of **X** and **Y** are both doubled? [1]

(b) The table below shows the initial concentrations of two compounds, **A** and **B**, and also the initial rate of the reaction that takes place between them at constant temperature.

Experiment	*[A]/mol dm^{-3}*	*[B]/mol dm^{-3}*	*Initial rate/mol dm^{-3} s^{-1}*
1	0.2	0.2	3.5×10^{-4}
2	0.4	0.4	1.4×10^{-3}
3	0.8	0.4	5.6×10^{-3}

(i) Determine the overall order of the reaction between **A** and **B**. Explain how you reached your conclusion. [2]

(ii) Determine the order of reaction with respect to compound **B**. Explain how you reached your conclusion. [2]

(iii) Write the rate equation for the overall reaction. [1]

(iv) Calculate the value of the rate constant, stating its units. [2]

[Total: 11]

Assessment and Qualifications Alliance Unit 4 Specimen Test Q2 2000

Chapter 2

Chemical equilibrium

The following topics are covered in this chapter:

- *The equilibrium constant K_c*
- *The equilibrium constant K_p*
- *Acids and bases*
- *The pH scale*
- *pH changes*

Key points from AS

- **Chemical equilibrium**
  Revise AS pages 104–108

During the study of equilibrium in AS Chemistry, the concept of dynamic equilibrium is introduced and le Chatelier's principle is used to predict how a change in conditions may alter the equilibrium position. For A2 Chemistry, you will build upon this knowledge and understanding to find out the exact position of equilibrium using the Equilibrium Law.

2.1 The equilibrium constant, K_c

After studying this section you should be able to:

- *deduce, for homogeneous reactions, K_c in terms of concentrations*
- *calculate values of K_c given appropriate data*
- *understand that K_c is changed only by changes in temperature*
- *understand how the magnitude of K_c relates to the equilibrium position*
- *calculate, from data, the concentrations present at equilibrium*

LEARNING SUMMARY

The equilibrium law

AQA	M4	SALTERS	M4
EDEXCEL	M4	WJEC	CH5
OCR	M6	NICCEA	M4
NUFFIELD	M4		

The equilibrium law states that, for an equation:

$$a\,\mathbf{A} + b\,\mathbf{B} \rightleftharpoons c\,\mathbf{C} + d\,\mathbf{D},$$

$$K_c = \frac{[\mathbf{C}]^c\,[\mathbf{D}]^d}{[\mathbf{A}]^a\,[\mathbf{B}]^b}$$

- $[\mathbf{A}]^a$, etc., are the **equilibrium** concentrations of the reactants and products of the reaction.
- Each product and reactant has its equilibrium concentration raised to the **power** of its **balancing number** in the equation.
- The equilibrium concentrations of the **products** are multiplied together on **top** of the fraction.
- The equilibrium concentrations of the **reactants** are multiplied together on the **bottom** of the fraction.

The overall equation for the reaction is used to write K_c.

Working out K_c

For the equilibrium: $H_2(g) + I_2(g) \rightleftharpoons 2HI(g)$

applying the equilibrium law above: $K_c = \dfrac{[HI(g)]^2}{[H_2(g)]\,[I_2(g)]}$

Equilibrium concentrations of $H_2(g)$, $I_2(g)$ and $HI(g)$ are shown below:

$[H_2(g)]$ /mol dm^{-3}	$[I_2(g)]$ /mol dm^{-3}	$[HI(g)]$ /mol dm^{-3}
0.0114	0.00120	0.0252

$$K_c = \frac{[HI(g)]^2}{[H_2(g)]\,[I_2(g)]} = \frac{0.0252^2}{0.0114 \times 0.00120} = 46.4$$

For A2 Chemistry, most courses only use **homogeneous equilibria** – all species are in the same phase:
all gaseous (g)
or
all aqueous (aq)
or
all liquid (l).

The units must be worked out afresh for each equilibrium.

See also p. 21 in which the units of the rate constant are discussed.

Units of K_c

- In the K_c expression, each concentration value is replaced by its units:

$$K_c = \frac{[HI(g)]^2}{[H_2(g)]\,[I_2(g)]} = \frac{(mol\ dm^{-3})^2}{(mol\ dm^{-3})\ (mol\ dm^{-3})}$$

- For this equilibrium, the units cancel.

∴ in the equilibrium: $H_2(g) + I_2(g) \rightleftharpoons 2HI(g)$, K_c has no units.

Properties of K_c

The magnitude of K_c indicates the extent of a chemical equilibrium.

- $K_c = 1$ indicates an equilibrium halfway between reactants and products.
- $K_c = 100$ indicates an equilibrium well in favour of the products.
- $K_c = 1 \times 10^{-2}$ indicates an equilibrium well in favour of the reactants.

K_c indicates how FAR a reaction proceeds; **not** how FAST.
K_c is unaffected by changes in concentration or pressure.
K_c can **only** be changed by altering the temperature.

Although changes in K_c affect the equilibrium yield, the actual conditions used may be different.

In an exothermic reaction, K_c increases and the equilibrium yield increases as temperature decreases.

However, a decrease in temperature may slow down the reaction so much that the reaction is stopped.

In practice, a compromise will be needed where equilibrium and rate are considered together – a reasonable equilibrium yield must be obtained in a reasonable length of time.

How does K_c vary with temperature?

It cannot be stressed too strongly that K_c can be changed only by altering the temperature. How K_c changes depends upon whether the reaction gives out or takes in heat energy.

In an **exothermic** reaction, K_c **decreases** with increasing temperature. Raising the temperature reduces the equilibrium yield of products.

$$H_2(g) + I_2(g) \rightleftharpoons 2HI(g): \qquad \Delta H^{\ominus}_{298} = -9.6\ kJ\ mol^{-1}$$

temperature /K	500	700	1100
K_c	160	54	25

In an **endothermic** reaction, K_c **increases** with increasing temperature. Raising the temperature increases the equilibrium yield of products.

$$N_2(g) + O_2(g) \rightleftharpoons 2NO(g): \qquad \Delta H^{\ominus}_{298} = +180\ kJ\ mol^{-1}$$

temperature /K	500	700	1100
K_c	5×10^{-13}	4×10^{-8}	1×10^{-5}

Progress check

1 For each of the following equilibria, write down the expression for K_c. State the units for K_c for each reaction.
(a) $N_2O_4(g) \rightleftharpoons 2NO_2(g)$
(b) $CO(g) + 2H_2(g) \rightleftharpoons CH_3OH(g)$
(c) $H_2(g) + Br_2(g) \rightleftharpoons 2HBr(g)$
(d) $2SO_2(g) + O_2(g) \rightleftharpoons 2SO_3(g)$

2 Explain whether the two reactions, **A** and **B**, are exothermic or endothermic.

temperature /K	*numerical value of K_c*	
	*reaction **A***	*reaction **B***
200	5.51×10^{-8}	4.39×10^{4}
400	1.46	4.03
600	3.62×10^{2}	3.00×10^{-2}

1 (a) $K_c = \frac{[NO_2(g)]^2}{[N_2O_4(g)]}$ (b) $K_c = \frac{[CH_3OH(g)]}{[CO(g)]\,[H_2(g)]^2}$ (c) $K_c = \frac{[HBr(g)]^2}{[H_2(g)]\,[Br_2(g)]}$ (d) $K_c = \frac{[SO_3(g)]^2}{[SO_2(g)]^2\,[O_2(g)]}$

2 Reaction **A**: endothermic; K_c increases with increasing temperature.
Reaction **B**: exothermic; K_c decreases with increasing temperature.

Determination of K_c from experiment

AQA	M4	SALTERS	M4
EDEXCEL	M4	WJEC	CH5
OCR	M6	NICCEA	M4
NUFFIELD	M4		

The equilibrium constant K_c can be calculated using experimental results. The example below shows how to:

- determine the equilibrium concentrations of the components in an equilibrium mixture
- calculate K_c.

The ethyl ethanoate esterification equilibrium

The most important stage in this calculation is to find the **change** in the number of moles of each species in the equilibrium.

0.200 mol CH_3COOH and 0.100 mol C_2H_5OH were mixed together with a trace of acid catalyst in a total volume of 250 cm^3. The mixture was allowed to reach equilibrium:

$$CH_3COOH + C_2H_5OH \rightleftharpoons CH_3COOC_2H_5 + H_2O.$$

Analysis of the mixture showed that 0.115 mol of CH_3COOH were present at equilibrium.

First summarise the results

It is useful to summarise the results as an 'I.C.E.' table: '**I**nitial/**C**hange/**E**quilibrium'.

	CH_3COOH	C_2H_5OH	$CH_3COOC_2H_5$	H_2O
Initial no. of moles	0.200	0.100	0	0
Change in moles				
Equilibrium no. of moles	0.115			

From the CH_3COOH values:

- moles of CH_3COOH that reacted = 0.200 – 0.115 = **0.085** mol

Find the change in moles of each component in the equilibrium

The amount of each component can be determined from the balanced equation. The change in moles of CH_3COOH is already known.

In this experiment, 0.085 mol CH_3COOH has **reacted** with 0.085 mol C_2H_5OH to form 0.085 mol $CH_3COOC_2H_5$ and 0.085 mol H_2O.

equation:	CH_3COOH +	C_2H_5OH	$\rightleftharpoons$	$CH_3COOC_2H_5$ +	H_2O
molar quantities:	1 mol	1 mol	$\longrightarrow$	1 mol	1 mol
change/mol:	**–0.085**	–0.085		+0.085	+0.085

Equilibrium concentrations are determined for each component

Note that the total volume is 250 cm^3 (0.250 dm^3). The concentration must be expressed as mol dm^{-3}.

	CH_3COOH +	C_2H_5OH	$\rightleftharpoons$	$CH_3COOC_2H_5$ +	H_2O
Initial amount/mol	0.200	0.100		0	0
Change in moles	–0.085	–0.085		+0.085	+0.085
Equilibrium amount/mol	0.115	0.015		0.085	0.085
Equilibrium conc. /mol dm^{-3}	$\frac{0.115}{0.250}$	$\frac{0.015}{0.250}$		$\frac{0.085}{0.250}$	$\frac{0.085}{0.250}$

Write the expression for K_c and substitute values

$$K_c = \frac{[CH_3COOC_2H_5]\,[H_2O]}{[CH_3COOH]\,[C_2H_5OH]} = \frac{\frac{0.085}{0.250} \times \frac{0.085}{0.250}}{\frac{0.115}{0.250} \times \frac{0.015}{0.250}}$$

Calculate K_c

$\therefore$ K_c = 4.19 (no units: all units cancel)

Progress check

1 Several experiments were set up for the $H_2(g)$, $I_2(g)$ and $HI(g)$ equilibrium. The equilibrium concentrations are shown below.

$[H_2(g)]$ /mol dm^{-3}	$[I_2(g)]$ /mol dm^{-3}	$[HI(g)]$ /mol dm^{-3}
0.0092	0.0020	0.0296
0.0077	0.0031	0.0334
0.0092	0.0022	0.0308
0.0035	0.0035	0.0235

Calculate the value for K_c for each experiment. Hence show that each experiment has the same value for K_c (allowing for experimental error). Work out an average value for K_c.

2 2 moles of ethanoic acid, CH_3COOH were mixed with 3 moles of ethanol and the mixture allowed to reach equilibrium.

$$CH_3COOH + C_2H_5OH \rightleftharpoons CH_3COOC_2H_5 + H_2O$$

At equilibrium, 0.5 moles of ethanoic acid remained.

(a) Work out the equilibrium concentrations of each component in the mixture. (You will need to use V to represent the volume but this will cancel out in your calculation.)

(b) Use these values to calculate K_c.

3 When 0.50 moles of $H_2(g)$ and 0.18 moles of $I_2(g)$ were heated at 500°C, the equilibrium mixture contained 0.01 moles of $I_2(g)$.

The equation is:

$H_2(g) + I_2(g) \rightleftharpoons 2HI(g)$

(a) How many moles of $I_2(g)$ reacted?

(b) How many moles of $H_2(g)$ were present at equilibrium?

(c) How many moles of $HI(g)$ were present at equilibrium?

(d) Calculate the equilibrium constant, K_c for this reaction.

1 values: 47.6; 46.7; 46.9; 45.1. average = 46.6

2 (a) CH_3COOH, 0.5/V mol dm^{-3}; C_2H_5OH, 1.5/V mol dm^{-3}; $CH_3COOC_2H_5$, 1.5/V mol dm^{-3}; H_2O, 1.5/V mol dm^{-3}.
(b) K_c = 3.

3 (a) 0.17 mol (b) 0.33 mol (c) 0.34 mol (d) 35.

2.2 The equilibrium constant, K_p

LEARNING SUMMARY

After studying this section you should be able to:

- *understand and use the terms mole fraction and partial pressure*
- *calculate from data the partial pressures present at equilibrium*
- *deduce, for homogeneous reactions, K_p in terms of partial pressures*
- *calculate from data the value of K_p, including determination of units*

Partial pressure

AQA	M4	SALTERS	M5
EDEXCEL	M4	WJEC	CH5
OCR	M6	NICCEA	M4
NUFFIELD	M4		

Equilibria involving gases are usually expressed in terms of K_p, the equilibrium constant in terms of partial pressures.

KEY POINT

In a gas mixture, the *partial pressure* of a gas, p, is the contribution that a gas makes towards the total pressure, P.

The air is a gas mixture with approximate molar proportions of 80% N_2(g) and 20% O_2(g).

- The partial pressure of N_2(g) is 80% of the total pressure.
- The partial pressure of O_2(g) is 20% of the total pressure.

The mole fraction of a gas is the same as its proportion by volume.

As with all fractions, the sum of the mole fractions in a mixture must equal ONE.

The sum of the partial pressures in a mixture must equal the total pressure.

KEY POINT

For a gas **A** in a gas mixture:

$$\text{the mole fraction of } \mathbf{A},\ x_A = \frac{\text{number of moles of } \mathbf{A}}{\text{total number of moles in gas mixture}}$$

partial pressure of **A**, p_A = mole fraction of **A** × total pressure = $x_A \times P$

The mole fractions in air are: $x_{N_2} = \frac{80}{100} = 0.8$; $x_{O_2} = \frac{20}{100} = 0.2$

At normal atmospheric pressure, 100 kPa,

$p_{N_2} = 0.8 \times 100 = 80$ kPa; $p_{O_2} = 0.2 \times 100 = 20$ kPa

Progress check

1 A gas mixture at a total pressure of 300 kPa contains 3 moles of N_2(g) and 1 mole of O_2(g).

(a) What is the mole fraction of each gas?

(b) What is the partial pressure of each gas?

1 (a) N_2, 0.75; O_2, 0.25. (b) N_2, 225 kPa; O_2, 75 kPa.

The equilibrium constant, K_p

AQA	M4	SALTERS	M5
EDEXCEL	M4	WJEC	CH5
OCR	M6	NICCEA	M4
NUFFIELD	M4		

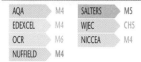

K_p is written in a similar way to K_c but with partial pressures replacing concentration terms.

For the equilibrium: $H_2(g) + I_2(g) \rightleftharpoons 2HI(g)$

the equilibrium constant in terms of partial pressures, K_p, is given by:

$$K_p = \frac{p_{HI}^{\,2}}{p_{H_2} \times p_{I_2}}$$

Don't use [ ] in K_p. This is a common exam mistake.

K_p only includes gases. Ignore any other species.

- p means the equilibrium partial pressure
- suitable units for partial pressures are kilopascals (kPa) or pascals (Pa) – but the same unit must used for all gases

- the power to which the partial pressures is raised is the *balancing number* in the chemical equation.

Working out K_p

An equilibrium mixture contains 13.5 mol $N_2(g)$, 3.6 mol $H_2(g)$, and 1.0 mol $NH_3(g)$. The total equilibrium pressure is 200 kPa. Calculate K_p.

Find the mole fractions and partial pressures

Total number of gas moles = 13.5 + 3.6 + 1.0 = 18.1 mol

Check that the partial pressures add up to give the total pressure: 149 + 40 + 11 = 200 kPa

$$p_{N_2} = \frac{13.5}{18.1} \times 200 = 149 \text{ kPa} \qquad p_{H_2} = \frac{3.6}{18.1} \times 200 = 40 \text{ kPa}$$

$$p_{NH_3} = \frac{1.0}{18.1} \times 200 = 11 \text{ kPa}$$

Calculate K_p

For the reaction: $N_2(g) + 3H_2(g) \rightleftharpoons 2NH_3(g)$,

$$K_p = \frac{{p_{NH_3}}^2}{p_{N_2} \times {p_{H_2}}^3}$$

$$\therefore\ K_p = \frac{11^2}{149 \times 40^3} = 1.27 \times 10^{-5} \text{ kPa}^{-2}$$

The units for K_p are found by replacing each partial pressure value in the K_p expression by its units:

substituting units: $K_p = \dfrac{(\text{kPa})^2}{(\text{kPa})\,(\text{kPa})^3}$ $\quad\therefore$ units of K_p are: kPa^{-2}

Progress check

1 For the following equilibria, write an expression for K_p and work out the value for K_p including units.
(a) $2HI(g) \rightleftharpoons H_2(g) + I_2(g)$
partial pressures: HI(g), 56 kPa; $H_2(g)$, 22 kPa; $I_2(g)$, 22 kPa
(b) $2NO_2(g) \rightleftharpoons 2NO(g) + O_2(g)$
partial pressures: $NO_2(g)$, 45 kPa; NO(g), 60 kPa; $O_2(g)$, 30kPa

2 In the equilibrium: $2SO_2(g) + O_2(g) \rightleftharpoons 2SO_3(g)$
2.0 mol of $SO_2(g)$ were mixed with 1.0 mol $O_2(g)$. The mixture was allowed to reach equilibrium at constant temperature and a constant pressure of 900 kPa in the presence of a catalyst. At equilibrium, 0.5 mol of the $SO_2(g)$ had reacted.
(a) How many moles of SO_2, O_2 and SO_3 were in the equilibrium mixture?
(b) What were the mole fractions and partial pressures of SO_2, O_2 and SO_3 in the equilibrium mixture?
(c) Calculate K_p

1 (a) 0.15 (b) 53 kPa.
2 (a) 1.5 mol SO_2; 0.75 mol O_2; 0.5 mol SO_3.
(b) $x(SO_2) = 0.55$; $x(O_2) = 0.27$; $x(SO_3) = 0.18$.
$p(SO_2) = 495$ kPa; $p(O_2) = 243$ kPa; $p(SO_3) = 162$ kPa.
(c) 4.4×10^{-4} kPa^{-1}

2.3 Acids and bases

After studying this section you should be able to:

LEARNING SUMMARY

- *describe what is meant by Brønsted–Lowry acids and bases*
- *understand conjugate acid–base pairs*
- *understand the difference between a strong and weak acid*
- *define the acid dissociation constant, K_a*

Key point from AS

- **Calculations in acid–base titrations**
  Revise AS page 40
- **Acids and bases**
  Revise AS pages 109–110

During the study of acids and bases at GCSE, you learnt that the pH scale can be used to measure the strength of acids and bases. You also learnt the reactions of acids with metals, carbonates and alkalis. During AS Chemistry, you revisited these reactions and found out how titrations can be used to measure the concentration of an unknown acid or base. For A2 Chemistry, you will study acids in terms of proton transfer, the strength of acids, pH and buffers.

Brønsted–Lowry acids and bases

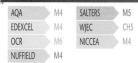

AQA	M4	SALTERS	M5
EDEXCEL	M4	WJEC	CH5
OCR	M6	NICCEA	M4
NUFFIELD	M4		

Key points from AS

- **Acids, bases and alkalis**
  Revise AS pages 109–110

An acid is a proton donor.

A base is a proton acceptor.

An acid–base reaction involves proton, H^+, transfer. This idea was first developed by Brønsted and Lowry.

KEY POINT

- A Brønsted–Lowry **acid** is a proton donor.
- A Brønsted–Lowry **base** is a proton acceptor.
- An **alkali** is a base that dissolves in water forming $OH^-(aq)$ ions.

This means that a molecule of an acid contains a hydrogen atom that can be released as a positive hydrogen ion or proton, H^+.

Acid–base pairs

Acids and bases are linked by H^+ as **conjugate pairs**:

- the **conjugate acid** donates H^+
- the **conjugate base** accepts H^+.

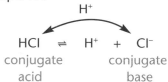

$$\underset{\text{conjugate acid}}{HCl} \rightleftharpoons H^+ + \underset{\text{conjugate base}}{Cl^-}$$

(H^+ transferred between the conjugate acid and conjugate base)

Examples of some conjugate acid–base pairs are shown below.

	acid				base
hydrochloric acid	HCl	$\rightleftharpoons$	H^+	+	Cl^-
sulphuric acid	H_2SO_4	$\rightleftharpoons$	H^+	+	HSO_4^-
ethanoic acid	CH_3COOH	$\rightleftharpoons$	H^+	+	CH_3COO^-

An acid needs a base

An acid can only donate a proton if there is a base to accept it. Most reactions of acids take place in aqueous conditions with water acting as the base. By mixing an acid with a base, an equilibrium is set up comprising **two** acid–base **conjugate pairs**.

The equilibrium system in aqueous hydrochloric acid is shown below.

		proton donor	←— H^+ —→	proton acceptor		
		ACID 1				BASE 1
$H_2O(l)$	+	$HCl(aq)$	⇌	$H_3O^+(aq)$	+	$Cl^-(aq)$
BASE 2				ACID 2		
proton acceptor	←— H^+ —→			proton donor		

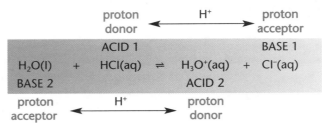

You should be able to identify acid–base pairs in equations such as this.

- In a reaction involving an aqueous acid such as hydrochloric acid, the 'active' species is the **oxonium** ion, H_3O^+ (ACID 2 below).
- Formation of the oxonium ion requires **both** an acid **and** water.

In equations, the oxonium ion H_3O^+ is usually shown as $H^+(aq)$.

Progress check

1 Identify the acid–base pairs in the acid–base equilibria below.

(a) $HNO_3 + H_2O \rightleftharpoons H_3O^+ + NO_3^-$

(b) $NH_3 + H_2O \rightleftharpoons NH_4^+ + OH^-$

(c) $H_2SO_4 + H_2O \rightleftharpoons HSO_4^- + H_3O^+$

2 Write equations for the following acid–base equilibria:

(a) hydrochloric acid and hydroxide ions

(b) ethanoic acid, CH_3COOH, and water.

1 (a) acid 1: HNO_3, base 1: NO_3^-; acid 2: H_3O^+, base 2: H_2O

(b) acid 1: H_2O, base 1: OH^-; acid 2: NH_4^+, base 2: NH_3

(c) acid 1: H_2SO_4, base 1: HSO_4^-; acid 2: H_3O^+, base 2: H_2O

2 (a) $HCl + OH^- \rightleftharpoons H_2O + Cl^-$

(b) $CH_3COOH + H_2O \rightleftharpoons H_3O^+ + CH_3COO^-$

Strength of acids and bases

AQA	M4	SALTERS	M5
EDEXCEL	M4	WJEC	CH5
OCR	M6	NICCEA	M4
NUFFIELD	M4		

The acid–base equilibrium of an acid, HA, in water is shown below.

$$HA(aq) + H_2O(l) \rightleftharpoons H_3O^+(aq) + A^-(aq)$$

To emphasise the loss of H^+, this can be shown more simply as **dissociation** of the acid HA:

$$HA(aq) \rightleftharpoons H^+(aq) + A^-(aq).$$

The **strength** of an acid shows the extent of dissociation into H^+ and A^-.

A strong acid is completely dissociated.

Strong acids

A **strong** acid, such as nitric acid, HNO_3, is a **good** proton donor.

- The equilibrium position lies well over to the right.
- There is almost **complete** dissociation.

$$\xrightarrow{\text{equilibrium}}$$

$$HNO_3(aq) \rightleftharpoons H^+(aq) + NO_3^-(aq)$$

- Virtually all of the potential acidic power has been released as $H^+(aq)$.
- At equilibrium, $[H^+(aq)]$ is much greater than $[HNO_3(aq)]$.

A weak acid only partially dissociates.

Weak acids

A **weak** acid, such as ethanoic acid, CH_3COOH, is a **poor** proton donor.

- The equilibrium position lies well over to the left.
- There is only **partial** dissociation.

$$\xleftarrow{\text{equilibrium}}$$

$$CH_3COOH(aq) \rightleftharpoons H^+(aq) + CH_3COO^-(aq)$$

Only a small proportion of the potential acidic power has been released as $H^+(aq)$. At equilibrium, $[CH_3COOH(aq)]$ is much greater than $[H^+(aq)]$.

K_a is just a special equilibrium constant K_c for equilibria showing the dissociation of acids.

[HA(aq)], [H$^+$(aq)] and [A$^-$(aq)] are equilibrium concentrations.

The acid dissociation constant, K_a

The extent of acid dissociation is shown by an equilibrium constant called the **acid dissociation constant**, K_a.

For the reaction: $HA(aq) \rightleftharpoons H^+(aq) + A^-(aq)$,

$$K_a = \frac{[H^+(aq)]\,[A^-(aq)]}{[HA(aq)]}$$

units: $K_a = \dfrac{(\text{mol dm}^{-3})^2}{(\text{mol dm}^{-3})} = \text{mol dm}^{-3}$.

- A **large** K_a value shows that the extent of **dissociation is large** – the acid is strong.
- A **small** K_a value shows that the extent of **dissociation is small** – the acid is weak.

Acid strength and concentration

Concentrated and *dilute* are terms used to describe the **amount** of dissolved acid in a solution.

Strong and *weak* are terms used to describe the degree of **dissociation** of an acid.

The distinction between the strength and concentration of an acid is important.

KEY POINT

Concentration is the **amount** of an acid dissolved in 1 dm^3 of solution.
- Concentration is measured in mol dm^{-3}.

Strength is the extent of **dissociation** of an acid.
- Strength is measured as K_a in units determined from the equilibrium equation.

Progress check

1 For each of the following acid–base equilibria, write down the expression for K_a. State the units of K_a for each reaction.
(a) $HCOOH(aq) \rightleftharpoons H^+(aq) + HCOO^-(aq)$
(b) $C_6H_5COOH(aq) \rightleftharpoons H^+(aq) + C_6H_5COO^-(aq)$

2 Samples of two acids, hydrochloric acid and ethanoic acid, have the same concentration: 0.1 mol dm^{-3}. Explain why one 'dilute acid' is *strong* whereas the other 'dilute acid' is *weak*!!

1 (a) $K_a = \dfrac{[H^+(aq)]\,[HCOO^-(aq)]}{[HCOOH(aq)]}$ units: mol dm^{-3}
(b) $K_a = \dfrac{[H^+(aq)]\,[C_6H_5COO^-(aq)]}{[C_6H_5COOH(aq)]}$ units: mol dm^{-3}

2 Concentration applies to the amount, in mol, in 1 dm^3 of solution. Both solutions have 0.1 mol dissolved in 1 dm^3 of solution and are dilute. Hydrochloric acid is strong because its dissociation is near to complete. However, ethanoic acid only donates a small proportion of its potential protons, its dissociation is incomplete and it is a weak acid.

2.4 The pH scale

After studying this section you should be able to:

- *define the terms pH, pK_a and K_w*
- *calculate pH from $[H^+(aq)]$*
- *calculate $[H^+(aq)]$ from pH*
- *understand the meaning of the ionisation product of water, K_w*
- *calculate pH for strong bases*

LEARNING SUMMARY

pH and $[H^+(aq)]$

AQA	M4	SALTERS	M5
EDEXCEL	M4	WJEC	CH5
OCR	M6	NICCEA	M4
NUFFIELD	M4		

The pH scale	
pH	$[H^+]$ / mol dm^{-3}
0	1
1	1×10^{-1}
2	1×10^{-2}
3	1×10^{-3}
4	1×10^{-4}
5	1×10^{-5}
6	1×10^{-6}
7	1×10^{-7}
8	1×10^{-8}
9	1×10^{-9}
10	1×10^{-10}
11	1×10^{-11}
12	1×10^{-12}
13	1×10^{-13}
14	1×10^{-14}

The concentrations of $H^+(aq)$ ions in aqueous solutions vary widely between about 10 mol dm^{-3} and about 1×10^{-15} mol dm^{-3}.

The **pH scale** is a logarithmic scale used to overcome the problem of using this large range of numbers and to ease the use of negative powers.

pH is defined as: $pH = -\log_{10} [H^+(aq)]$
- where $[H^+(aq)]$ is the concentration of hydrogen ions in aqueous solution.
- $[H^+(aq)]$ can be calculated from pH using:
  $[H^+(aq)] = 10^{-pH}$

KEY POINT

Notice how the value of pH is linked to the power of 10.

pH	2	9	3.6	10.3
$[H^+(aq)]$/mol dm^{-3}	10^{-2}	10^{-9}	$10^{-3.6}$	$10^{-10.3}$

What does a pH value mean?

- A **low** value of $[H^+(aq)]$ matches a **high** value of pH.
- A **high** value of $[H^+(aq)]$ matches a **low** value of pH.
- A change of pH by 1 changes $[H^+(aq)]$ by 10 times.
- An acid of pH 4 contains 10 times the concentration of $H^+(aq)$ ions as an acid of pH 5.

Calculating the pH of strong acids

AQA	M4	SALTERS	M5
EDEXCEL	M4	WJEC	CH5
OCR	M6	NICCEA	M4
NUFFIELD	M4		

HA is **monoprotic**: each HA molecule can donate **one** H^+ ion.

10^x

This is the key to use on your calculator when doing pH and $[H^+]$ calculations.

For 10^X, press the **SHIFT** or **INV** key first.

For a strong acid, HA:
- we can assume complete dissociation
- the concentration of $H^+(aq)$ can be found directly from the acid concentration: $[H^+] = [HA]$.

You should be able to convert pH into $[H^+(aq)]$ and *vice versa*

Example 1

A strong acid, HA, has a concentration of 0.010 mol dm^{-3}. What is the pH?

Complete dissociation. $\therefore [H^+(aq)] = 0.010$ mol dm^{-3}

$$pH = -\log_{10} [H^+(aq)] = -\log_{10} (0.010) = \mathbf{2.0}$$

Example 2

A strong acid, HA, has a pH of 3.4. What is the concentration of $H^+(aq)$?

Complete dissociation. $\therefore [H^+(aq)] = 10^{-pH} = 10^{-3.4}$ mol dm^{-3}

$$\therefore [H^+(aq)] = \mathbf{3.98 \times 10^{-4}} \text{ mol dm}^{-3}$$

Hints for pH calculations

Calculations involving pH are easy once you have learnt how to use your calculator properly.

- Try the examples above until you can remember the order to press the keys.
- Try reversing each calculation to go back to the original value. Repeat several times until you have mastered how to use **your** calculator for pH calculations.
- **Don't** borrow a calculator or you will get confused. Different calculators may need the keys to be pressed in a different order!
- Look at your answer and decide whether it looks sensible.

KEY POINT

Learn: $pH = -\log_{10} [H^+(aq)]$
$[H^+(aq)] = 10^{-pH}$.

Progress check

1 Calculate the pH of solutions with the following $[H^+(aq)]$ values.
(a) 0.01 mol dm^{-3} (d) 2.50×10^{-3} mol dm^{-3}
(b) 0.0001 mol dm^{-3} (e) 8.10×10^{-6} mol dm^{-3}
(c) 1.0×10^{-13} mol dm^{-3} (f) 4.42×10^{-11} mol dm^{-3}

2 Calculate $[H^+(aq)]$ of solutions with the following pH values.
(a) pH 3 (c) pH 2.8 (e) pH 12.2
(b) pH 10 (d) pH 7.9 (f) pH 9.6

3 How many times more hydrogen ions are in an acid of pH 1 than an acid of pH 5?

4 How can a solution have a pH with a negative value?

1 (a) 2 (b) 4 (c) 13 (d) 2.60 (e) 5.09 (f) 10.4
2 (a) 1×10^{-3} mol dm^{-3} (d) 1.26×10^{-8} mol dm^{-3}
(b) 1×10^{-10} mol dm^{-3} (e) 6.31×10^{-13} mol dm^{-3}
(c) 1.58×10^{-3} mol dm^{-3} (f) 2.51×10^{-10} mol dm^{-3}
3 pH 1 has 10 000 times more H^+ ions than pH 5.
4 A solution with $[H^+(aq)] > 1$ mol dm^{-3} has a negative pH value.

Calculating the pH of weak acids

AQA	M4	SALTERS	M5
EDEXCEL	M4	WJEC	CH5
OCR	M6	NICCEA	M4
NUFFIELD	M4		

The pH of a weak acid HA can be calculated from:

- the **concentration** of the acid and
- the value of the acid dissociation constant, K_a.

Assumptions and approximations

Consider the equilibrium of a weak aqueous acid HA(aq):

$$HA(aq) \rightleftharpoons H^+(aq) + A^-(aq)$$

- Assuming that only a very small proportion of HA dissociates, the equilibrium concentration of HA(aq) will be very nearly the same as the concentration of undissociated HA(aq).

$$\therefore [HA(aq)]_{equilibrium} \approx [HA(aq)]_{start}$$

- Assuming that there is a negligible proportion of $H^+(aq)$ from ionisation of water:

$$[H^+(aq)] \approx [A^-(aq)]$$

- Using these approximations

$$K_a = \frac{[H^+(aq)]\,[A^-(aq)]}{[HA(aq)]} \qquad \therefore K_a \approx \frac{[H^+(aq)]^2}{[HA(aq)]}$$

For calculations, use $K_a \approx \dfrac{[H^+(aq)]^2}{[HA(aq)]}$

Take care to learn this method.
Many marks are dropped on exam papers by students who have not done so!

Example

For a weak acid [HA(aq)] = 0.100 mol dm^{-3}, $K_a = 1.70 \times 10^{-5}$ mol dm^{-3} at 25°C. Calculate the pH.

$$K_a = \frac{[H^+(aq)]\,[A^-(aq)]}{[HA(aq)]} \approx \frac{[H^+(aq)]^2}{[HA(aq)]}$$

$$\therefore\ 1.70 \times 10^{-5} = \frac{[H^+(aq)]^2}{0.100}$$

$$\therefore\ [H^+(aq)] = \sqrt{0.100 \times 1.70 \times 10^{-5}} = 0.00130 \text{ mol dm}^{-3}$$

$$pH = -\log_{10}[H^+(aq)] = -\log_{10}(0.00130) = \mathbf{2.89}$$

K_a and pK_a

AQA	M4	SALTERS	M5
EDEXCEL	M4	WJEC	CH5
OCR	M6	NICCEA	M4
NUFFIELD	M4		

Values of K_a can be made more manageable if expressed in a logarithmic form, pK_a (see also page 38: [H^+(aq)] and pH).

$$\mathbf{pK_a = -\log_{10}K_a}$$
$$\mathbf{K_a = 10^{-pKa}}$$

K_a and pK_a conversions are just like those between pH and H^+.

- A **low** value of K_a matches a **high** value of pK_a
- A **high** value of K_a matches a **low** value of pK_a

The smaller the pK_a value, the stronger the acid.

Comparison of K_a and pK_a

acid		*K_a / mol dm^{-3}*	*pK_a*
methanoic acid	HCOOH	1.6×10^{-4}	$-\log_{10}(1.6 \times 10^{-4})$ = **3.8**
benzoic acid	C_6H_5COOH	6.3×10^{-5}	$-\log_{10}(6.3 \times 10^{-5})$ = **4.2**

Progress check

1 Find the pH of solutions of a weak acid HA ($K_a = 1.70 \times 10^{-5}$ mol dm^{-3}), with the following concentrations:
(a) 1.00 mol dm^{-3}; (b) 0.250 mol dm^{-3}; (c) 3.50×10^{-2} mol dm^{-3}

2 Find values of K_a and pK_a for the following weak acids.
(a) 1.0 mol dm^{-3} solution with a pH of 4.5
(b) 0.1 mol dm^{-3} solution with a pH of 2.2
(c) 2.0 mol dm^{-3} solution with a pH of 3.8.

1 (a) 2.38 (b) 2.69 (c) 3.11
2 (a) $K_a = 1 \times 10^{-9}$ mol dm^{-3}, $pK_a = 9$
(b) K_a 3.98×10^{-4} mol dm^{-3}, $pK_a = 3.4$
(c) K_a 1.26×10^{-8} mol dm^{-3}, $pK_a = 7.9$

The ionisation of water and K_w

AQA	M4	SALTERS	M5
EDEXCEL	M4	WJEC	CH5
OCR	M6	NICCEA	M4
NUFFIELD	M4		

In water, a very small proportion of molecules dissociates into H^+(aq) and OH^-(aq) ions. The position of equilibrium lies well to the left of the equation below, representing this dissociation.

$$\xleftarrow{\text{equilibrium}}$$
$$H_2O(l) \rightleftharpoons H^+(aq) + OH^-(aq)$$

Treating water as a weak acid: $K_a = \dfrac{[H^+(aq)]\,[OH^-(aq)]}{[H_2O(l)]}$

Rearranging gives:

$$\underbrace{K_a \times [H_2O(l)]}_{\text{constant, } K_w} = [H^+(aq)]\,[OH^-(aq)]$$

$[H_2O(l)]$ is constant and is included within K_w

At 25°C,
$[H^+(aq)] \times [OH^-(aq)]$
$= 1 \times 10^{-14}$ mol^2 dm^{-6}.

KEY POINT

The constant K_w is called the **ionic product of water**
- $K_w = [H^+(aq)]\,[OH^-(aq)]$
- At 25°C, $K_w = 1.0 \times 10^{-14}$ mol^2 dm^{-6}.

In water, the concentrations of $H^+(aq)$ and $OH^-(aq)$ ions are the same.

- $[H^+(aq)] = [OH^-(aq)] = 10^{-7}$ mol dm^{-3} $(10^{-14} = 10^{-7} \times 10^{-7})$

Hydrogen ion and hydroxide ion concentrations

Linking $[H^+]$ and $[OH^-]$

$10^{-14} = [H^+]\,[OH^-]$

Water: pH = 7,
$[H^+] = 10^{-7}$ mol dm^{-3}
$10^{-14} = 10^{-7} \times 10^{-7}$

An acid: pH = 3
$[H^+] = 10^{-3}$ mol dm^{-3}
$10^{-14} = 10^{-3} \times 10^{-11}$

An alkali: pH = 10
$[H^+] = 10^{-10}$ mol dm^{-3}
$10^{-14} = 10^{-10} \times 10^{-4}$

All aqueous solutions contain $H^+(aq)$ and $OH^-(aq)$ ions. The proportions of these ions in a solution are determined by the pH.

In water	$[H^+(aq)] = [OH^-(aq)]$
In **acidic** solutions	$[H^+(aq)] > [OH^-(aq)]$
In **alkaline** solutions	$[H^+(aq)] < [OH^-(aq)]$

The concentrations of $H^+(aq)$ and $OH^-(aq)$ are linked by K_w.

- At 25°C $1.0 \times 10^{-14} = [H^+(aq)]\,[OH^-(aq)]$
- The indices of $[H^+(aq)]$ and $[OH^-(aq)]$ add up to –14.

Calculating the pH of strong alkalis

The pH of a strong alkali can be found using K_w.

For a strong alkali, BOH:

- we can assume complete dissociation
- the concentration of $OH^-(aq)$ can be found directly from the alkali concentration: $[BOH] = [OH^-]$.

To find the pH of an alkali, first find $[H^+]$ using K_w and $[OH^-]$.

Example

A strong alkali, BOH, has a concentration of 0.50 mol dm^{-3}.

What is the pH?

Complete dissociation. $\therefore [OH^-(aq)] = 0.50$ mol dm^{-3}

$$K_w = [H^+(aq)]\,[OH^-(aq)] = 1 \times 10^{-14} \text{ mol}^2 \text{ dm}^{-6}$$

$$\therefore [H^+(aq)] = \frac{K_w}{[OH^-(aq)]} = \frac{1 \times 10^{-14}}{0.50} = 2 \times 10^{-13} \text{ mol dm}^{-3}$$

$$pH = -\log_{10}[H^+(aq)] = -\log_{10}(2 \times 10^{-13}) = \mathbf{12.7}$$

Progress check

1 Find the $[H^+(aq)]$ and pH of the following alkalis at 25°C:
(a) 1×10^{-3} mol dm^{-3} $OH^-(aq)$
(b) 3.5×10^{-2} mol dm^{-3} $OH^-(aq)$.

2 Find the pH of the following solutions of strong bases at 25°C:
(a) 0.01 mol dm^{-3} KOH (aq)
(b) 0.20 mol dm^{-3} NaOH (aq).

1 (a) $[H^+(aq)] = 1 \times 10^{-11}$ mol dm^{-3}; pH = 11
(b) $[H^+(aq)] = 2.86 \times 10^{-13}$ mol dm^{-3}; pH = 12.5.
2 (a) pH = 12; (b) pH = 13.3.

2.5 pH changes

After studying this section you should be able to:

LEARNING SUMMARY

- *understand how an indicator changes colour at different pHs*
- *recognise the shapes of titration curves for acids and bases with different strengths*
- *explain the choice of suitable indicators for acid–base titrations*
- *state what is meant by a buffer solution*
- *explain how pH is controlled by each component in a buffer solution*
- *calculate the pH of a buffer solution*

Indicators and end-points

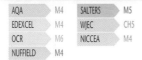

AQA	M4	SALTERS	M5
EDEXCEL	M4	WJEC	CH5
OCR	M6	NICCEA	M4
NUFFIELD	M4		

Indicators as weak acids

An acid–base indicator is a weak acid, represented simply as HInd.

- The weak acid, HInd, and its conjugate base, Ind^-, have different colours.

E.g. for methyl orange:

RED				YELLOW
HInd(aq)	⇌	H^+(aq)	+	Ind^-(aq)
weak acid				conjugate base

At the end-point of a titration:

- **HInd** and **Ind^-** are present in **equal** concentrations.

When using methyl orange as indicator:

- the colour at the end-point is orange, from equal concentrations of the weak acid HInd (red) and its conjugate base Ind^- (yellow)
- the pH of the end-point is called the pK_{ind} of the indicator.

Each indicator has its own pK_{ind} value at which its colour changes.

pH ranges for common indicators

An indicator changes colour over a range of about 2 pH units within which is the pK_{ind} value of the indicator. The pH range for the indicators methyl orange and phenolphthalein are shown below.

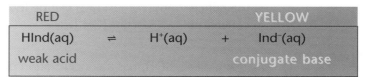

pH 0 1 2 3 4 5 6 7 8 9 10 11 12 13 14
red ⟷ yellow (pH 2–4) methyl orange, pK_{ind} = 3.7
colourless ⟷ pink (pH 8–10) phenolphthalein, pK_{ind} = 9.3

Titration curves

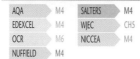

AQA	M4	SALTERS	M4
EDEXCEL	M4	WJEC	CH5
OCR	M6	NICCEA	M4
NUFFIELD	M4		

Choosing an indicator using titration curves

A titration curve shows the changes in pH during a titration.
In the titration curves on the next page:

- different combinations of strong and weak acids have been used
- the pK_{ind} values are shown for the indicators methyl orange (**MO**) and phenolphthalein (**P**).

strong acid/strong alkali

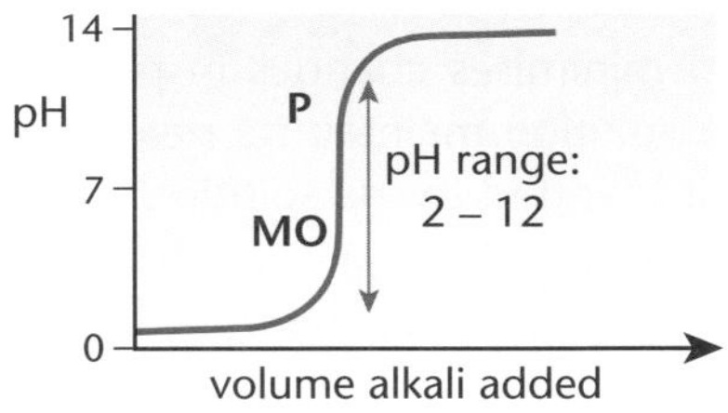

both indicators suitable

strong acid/weak alkali

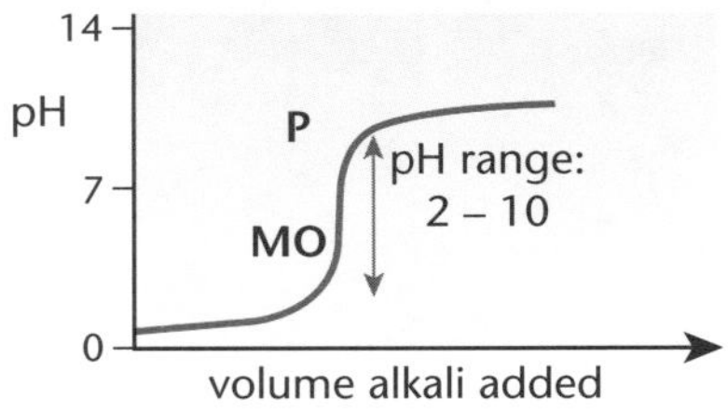

methyl orange (**MO**) *suitable*
phenolphthalein (**P**) *unsuitable*

weak acid/strong alkali

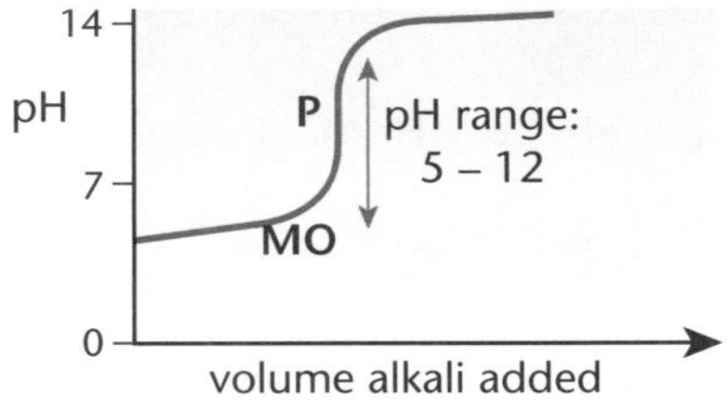

phenolphthalein (**P**) *suitable*
methyl orange (**MO**) *unsuitable*

weak acid/weak alkali

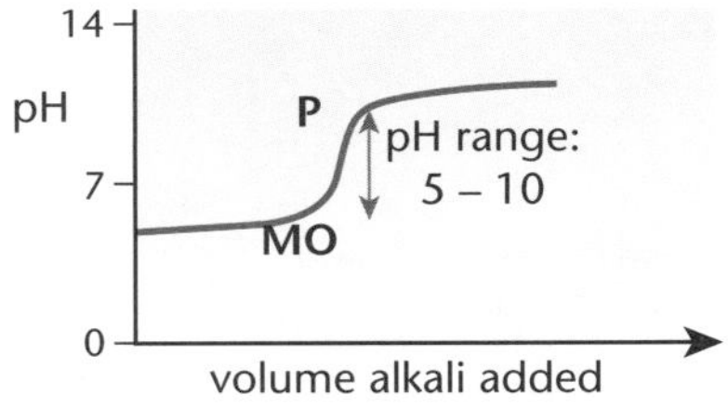

neither indicator suitable

Choose the correct indicator

Strong acid/strong alkali
phenolphthalein ✓
methyl orange ✓

Strong acid/weak alkali
phenolphthalein ✗
methyl orange ✓

Weak acid/strong alkali
phenolphthalein ✓
methyl orange ✗

Weak acid/weak alkali
phenolphthalein ✗
methyl orange ✗

Key features of titration curves

- The pH changes rapidly at the near vertical portion of the titration curve. This is the **end-point** or **equivalence point** of the titration.
- The sharp change in pH is brought about by a very small addition of alkali, typically the addition of one drop.
- The indicator is only suitable if its pK_{ind} value is within the pH range of the near vertical portion of the titration curve.

Titrations of weak acids/weak alkalis

- The pH changes slowly through the end point as alkali is added – there is **no** near vertical portion to the titration curve.
- The colour change of the indicator is gradual.
- It is very difficult to see a sharp colour change.
- The pH change is often followed using a pH meter instead of an indicator.

Progress check

1 The indicator, bromophenol blue, has an HInd form that is a yellow colour and an Ind^- form that is a blue colour.
(a) Write an equation to link these two forms of the indicator.
(b) Explain, with reasons, the colour of bromophenol blue
  (i) in a strong acid
  (ii) in a strong alkali
  (iii) at the end-point of a titration.

1 (a) $HInd \rightleftharpoons H^+ + Ind^-$
(b) (i) Bromophenol blue is yellow. The high concentration of $H^+(aq)$ moves the equilibrium to the left, forming HInd(aq).
(ii) Bromophenol blue is blue. The high concentration of $OH^-(aq)$ removes $H^+(aq)$ and moves the equilibrium to the right, forming $Ind^-(aq)$.
(iii) Bromophenol blue is green. At the end-point, there are equal concentrations of HInd and Ind^-, blue and yellow make green.

Buffer solutions

AQA	M4	SALTERS	M5
EDEXCEL	M4	WJEC	CH5
OCR	M6	NICCEA	M4
NUFFIELD	M4		

A buffer solution minimises changes in pH during the addition of an acid or an alkali. The buffer solution maintains a near constant pH by removing most of any acid or alkali that is added to the solution.

A **buffer solution** is a mixture of:

- a **weak acid**, HA, and
- its **conjugate base**, A^-:

$$\underset{\text{weak acid}}{HA(aq)} \rightleftharpoons H^+(aq) + \underset{\text{conjugate base}}{A^-(aq)}$$

In a buffer solution, the concentration of hydrogen ions, $[H^+(aq)]$, is very small compared with the concentrations of the weak acid $[HA(aq)]$ or the conjugate base $[A^-(aq)]$.

$$[H^+(aq)] << [HA(aq)] \quad \text{and} \quad [H^+(aq)] << [A^-(aq)].$$

How does a buffer act?

We can explain how a buffer minimises pH changes by using le Chatelier's principle.

Addition of an acid, $H^+(aq)$, to a buffer

On addition of an acid:

- $[H^+(aq)]$ is increased
- the pH change is opposed and the **equilibrium moves to the left, removing $[H^+(aq)]$** and forming HA(aq).
- the **conjugate base $A^-(aq)$** removes most of any added $[H^+(aq)]$.

A^- removes most of any added acid.

add acid, $H^+(aq)$ ↓

$$\underset{\text{weak acid}}{HA(aq)} \rightleftharpoons H^+(aq) + \underset{\text{conjugate base}}{A^-(aq)}$$

← **$H^+(aq)$ is removed**

Addition of an alkali, $OH^-(aq)$, to a buffer

This idea is very similar to that explaining the action of indicators, see p. 42.

On addition of an alkali:

- the added $OH^-(aq)$ reacts with the small concentration of $H^+(aq)$:

$$H^+(aq) + OH^-(aq) \longrightarrow H_2O(l)$$

- the pH change is opposed – the **equilibrium moves to the right, restoring $[H^+(aq)]$** as HA(aq) dissociates
- the **weak acid HA** restores most of any $[H^+(aq)]$ that has been removed.

add alkali, $OH^-(aq)$ ↑

$$\underset{\text{weak acid}}{HA(aq)} \rightleftharpoons H^+(aq) + \underset{\text{conjugate base}}{A^-(aq)}$$

→ **$H^+(aq)$ is restored**

HA removes added alkali.

KEY POINT

Although the two components in a buffer solution react with added acid and alkali, they **cannot stop** the pH from changing. They do however **minimise** pH changes.

Remember these buffers:
CH_3COOH/CH_3COONa
and
NH_4Cl/NH_3

Common buffer solutions

An acidic buffer

A common acidic buffer is an aqueous solution containing a mixture of ethanoic acid, CH_3COOH, and the ethanoate ion, CH_3COO^-.

- Ethanoic acid acts as the **weak acid**, CH_3COOH.
- Sodium ethanoate, $CH_3COO^-Na^+$, acts as a source of the **conjugate base**, CH_3COO^-.

An alkaline buffer

A common alkaline buffer is an aqueous solution containing a mixture of the ammonium ion, NH_4^+, and ammonia, NH_3.

- Ammonium chloride, $NH_4^+Cl^-$, acts as source of the **weak acid**, NH_4^+.
- Ammonia acts as the **conjugate base**, NH_3.

Calculations involving buffer solutions

AQA	M4	SALTERS	M5
EDEXCEL	M4	WJEC	CH5
OCR	M6	NICCEA	M4
NUFFIELD	M4		

For buffer calculations, learn:

$[H^+(aq)] = K_a \times \frac{[HA(aq)]}{[A^-(aq)]}$

The pH of a buffer can be controlled by the acid/base ratio.

The pH of a buffer solution depends upon:

- the acid dissociation constant, K_a, of the buffer system
- the **ratio** of the weak **acid** and its conjugate **base**.

For a buffer comprising the weak acid, HA, and its conjugate base, A^-,

$$K_a = \frac{[H^+(aq)]\,[A^-(aq)]}{[HA(aq)]}$$

$$\therefore [H^+(aq)] = K_a \times \frac{[HA(aq)]}{[A^-(aq)]}$$

acid dissociation constant (K_a) — **ratio** of the weak acid and its conjugate base ($\frac{[HA(aq)]}{[A^-(aq)]}$)

Example

Calculate the pH of a buffer comprising 0.30 mol dm^{-3} $CH_3COOH(aq)$ ($K_a = 1.7 \times 10^{-5}$ mol dm^{-3}) and 0.10 mol dm^{-3} $CH_3COO^-(aq)$.

$$[H^+(aq)] = K_a \times \frac{[HA(aq)]}{[A^-(aq)]}$$

$$\therefore [H^+(aq)] = 1.7 \times 10^{-5} \times \frac{0.30}{0.10} = 5.1 \times 10^{-5} \text{ mol dm}^{-3}$$

$$\therefore pH = -\log_{10} [H^+(aq)] = -\log_{10} (5.1 \times 10^{-5})$$

$\therefore$ pH of the buffer solution = **4.3**

Progress check

1 Three buffer solutions are made from benzoic acid, C_6H_5COOH and sodium benzoate, C_6H_5COONa with the following compositions:

Buffer A: 0.10 mol dm^{-3} C_6H_5COOH and 0.10 mol dm^{-3} C_6H_5COONa
Buffer B: 0.75 mol dm^{-3} C_6H_5COOH and 0.25 mol dm^{-3} C_6H_5COONa
Buffer C: 0.20 mol dm^{-3} C_6H_5COOH and 0.80 mol dm^{-3} C_6H_5COONa

(a) Write the equation for the equilibrium in these buffers.
(b) Write an expression for K_a of benzoic acid.
(c) Calculate the pH of each of the buffer solutions (K_a for $C_6H_5COOH = 6.3 \times 10^{-5}$ mol dm^{-3}.)

1 (a) $C_6H_5COOH \rightleftharpoons H^+(aq) + C_6H_5COO^-(aq)$
(b) $K_a = \frac{[H^+(aq)]\,[C_6H_5COO^-(aq)]}{[C_6H_5COOH(aq)]}$
(c) Buffer **A**: 4.2 Buffer **B**: 3.7 Buffer **C**: 4.8

Sample question and model answer

Make sure that your calculator skills are good and show your working. A correct method will secure most available marks.

Notice the use throughout of 3 significant figures (in the question and in answers).

These approximations simplify the K_a expression to:

$K_a \approx \frac{[H^+]^2}{[HA]}$

Notice that 3 marks means 3 key points.

Blood is an example of a buffered solution. Human blood is slightly basic and has a pH of approximately 7.40. If the pH falls, a condition known as acidosis can occur. Death may arise if the pH drops below 6.80.

(a) (i) What do you understand by the term a *buffered solution*?

A solution that resists a change in pH. ✓

(ii) Calculate the approximate hydrogen ion concentration, $[H^+(aq)]$, of normal human blood.

$[H^+] = 10^{-pH}$ ✓ $\therefore [H^+] = 10^{-7.40} = 3.98 \times 10^{-8}$ ✓ $mol\ dm^{-3}$

(iii) How many times greater is the hydrogen ion concentration in blood at pH 6.8 compared to blood at pH 7.4?

At pH 6.8, $[H^+] = 10^{-6.80} = 1.58 \times 10^{-7}$ ✓ $mol\ dm^{-3}$

$[H^+]$ is $\frac{1.58 \times 10^{-7}}{3.98 \times 10^{-8}}$ times greater ✓ = 3.97 times greater.

[5]

(b) Acidosis may occur as a result of strenuous exercise when glucose is converted into lactic acid, $CH_3CH(OH)CO_2H$. Lactic acid is a weak acid with an acid dissociation constant, K_a, of 8.40×10^{-4} mol dm^{-3}.

(i) Write an equation to show the dissociation of lactic acid into its ions.

$CH_3CH(OH)CO_2H \rightleftharpoons CH_3CH(OH)CO_2^- + H^+$ ✓

(ii) Write an expression for the acid dissociation constant, K_a, of lactic acid.

$$K_a = \frac{[CH_3CH(OH)CO_2^-(aq)]\,[H^+(aq)]}{[CH_3CH(OH)CO_2H(aq)]}$$ ✓

(iii) Calculate the pH of a 0.100 mol dm^{-3} solution of lactic acid.

$CH_3CH(OH)CO_2^- \approx H^+$

equilibrium conc of weak acid ≈ undissociated concentration of weak acid.

$$\therefore K_a \approx \frac{[H^+(aq)]^2}{[CH_3CH(OH)COOH(aq)]}$$

$$8.4 \times 10^{-4} = \frac{[H^+(aq)]^2}{[CH_3CH(OH)COOH(aq)]} = \frac{[H^+(aq)]^2}{0.100}$$ ✓

$[H^+(aq)] = \sqrt{0.100 \times 8.40 \times 10^{-4}} = 9.17 \times 10^{-3}$ ✓ $mol\ dm^{-3}$

pH = $-\log(9.17 \times 10^{-3}) = 2.04$. ✓

[5]

(c) Oxygen is carried in the blood attached to haemoglobin (*Hb*) in the form HbO_2 and is transported around the body. The haemoglobin is involved in a series of equilibria whose overall result can be represented by the following equation.

$$HbH^+(aq) + O_2(aq) \rightleftharpoons HbO_2(aq) + H^+(aq)$$

Show how an increase in the concentration of lactic acid influences this equilibrium and suggest a possible consequence of this within the body.

H^+ is added to right-hand side of equilibrium. ✓

The equilibrium moves to the left. ✓

This reduces the ability to transport oxygen. ✓

[3]

[Total: 13]

Cambridge How Far, How Fast? Q2 June 1996

Practice examination questions

1

Consider the esterification reaction:

$$CH_3COOH + CH_3CH_2OH \rightleftharpoons CH_3COOCH_2CH_3 + H_2O$$

(a) (i) Give the expression for K_c. [1]

(ii) An equilibrium mixture from the above reaction contains 0.90 mol of ethyl ethanoate, 0.90 mol of water, 0.10 mol of ethanoic acid and 2.1 mol of ethanol. It has a volume of 235 cm^3. Calculate K_c to 2 significant figures, having regard for the units. [3]

(iii) State, giving a reason, what would happen to the equilibrium composition if more ethanoic acid were to be added to the equilibrium mixture. [2]

(b) The reaction employs an acid catalyst, usually a little sulphuric acid. State, with a reason, the effect on the equilibrium position if the concentration of the acid catalyst were to be increased. [2]

[Total: 8]

London Module Test 2 Q1 (a)-(b) January 1999

2

In the Contact Process for the production of sulphuric acid, highly purified sulphur dioxide and oxygen react together to form sulphur trioxide. The process is carried out in the presence of vanadium(V) oxide, at about 700 K and at a pressure of 120 kPa according to the equation:

$$2SO_2(g) + O_2(g) \rightleftharpoons 2SO_3(g) \qquad \Delta H^\ominus = -190 \text{ kJ mol}^{-1}$$

Under these conditions, the partial pressures of sulphur dioxide and sulphur trioxide at equilibrium are 33 kPa and 39 kPa, respectively.

(a) What would be the effect on the yield of SO_3 of increasing the temperature? Explain your answer. [3]

(b) Determine the partial pressure and hence the mole fraction of oxygen in the equilibrium mixture. [3]

(c) (i) Write an expression for the equilibrium constant, K_p, for the reaction shown.

(ii) Calculate the value of the equilibrium constant, K_p, and state its units. [4]

[Total: 10]

NEAB Further Physical Chemistry Q3 June 1996

3

An equilibrium mixture of N_2O_4 and NO_2 appears as brown fumes.

$N_2O_4(g)$	$\rightleftharpoons$	$2NO_2(g)$
pale yellow		dark brown

A transparent glass syringe was filled with the gaseous mixture of N_2O_4 and NO_2 and its tip sealed. When the piston of the syringe was rapidly pushed well into the body of the syringe, thereby compressing the gas mixture considerably, the colour of the gas became momentarily darker but then became lighter again.

(a) Suggest why compressing the gases causes the mixture to darken. [1]

(b) Explain why the mixture turns lighter on standing. [2]

(c) Write an expression for the equilibrium constant, K_p, for this equilibrium. [1]

(d) 1.0 mole of N_2O_4 was allowed to reach equilibrium at 400 K. At equilibrium the partial pressure of N_2O_4 was found to be 0.15 atm. Given that the equilibrium constant K_p for this reaction is 48 atm, calculate the partial pressure of NO_2 in the equilibrium mixture. [3]

[Total: 7]

Edexcel Module Test 3 Q3(b) modified June 2000

Practice examination questions (continued)

4

(a) Define the term Brønsted–Lowry acid. [1]

(b) Write an equation for the reaction between gaseous hydrogen chloride and water. State the role of water in this reaction, using the Brønsted–Lowry definition. [2]

(c) Write an equation for the reaction between gaseous ammonia and water. State the role of water in this reaction, using the Brønsted–Lowry definition. [2]

(d) The ion $H_2NO_3^+$ is formed in the first stage of a reaction between concentrated nitric acid and an excess of concentrated sulphuric acid. In this first stage the two acids react in a 1 : 1 molar ratio. In the second stage, the $H_2NO_3^+$ ion decomposes to form the nitronium ion, NO_2^+. Write equations for these two reactions and state the role of nitric acid in the first reaction. [3]

(e) (i) Explain the term weak acid.
(ii) Write an expression for the acid dissociation constant, K_a of HA, a weak monoprotic acid.
(iii) The value of the acid dissociation constant for the monoprotic acid HX is 144 mol dm^{-3}. What does this suggest about the concentration of undissociated HX in dilute aqueous solution?
(iv) State whether HX should be classified as a strong acid or a weak acid. Justify your answer. [5]

[Total:13]

NEAB Equilibria and Inorganic Chemistry Q1 June 2000

5

(a) (i) Define pH.
(ii) Define the term 'weak acid' as applied to methanoic acid, HCOOH. [3]

(b) Calculate the pH of the following solutions:
(The ionic product of water, $K_w = 1.00 \times 10^{-14}$ mol^2 dm^{-6} at 25°C.
The acid dissociation constant for methanoic acid is 1.78×10^{-4} mol dm^{-3}.)
(i) a solution of hydrochloric acid of concentration 0.152 mol dm^{-3}
(ii) a solution of sodium hydroxide of concentration 0.747 mol dm^{-3}
(iii) a solution of methanoic acid of concentration 0.152 mol dm^{-3}. [9]

(c) (i) What is the principle property of a buffer solution?
(ii) The acid dissociation constant, K_a for ethanoic acid is 1.80×10^{-5} mol dm^{-3}. Calculate the pH of a buffer solution which has a concentration of 0.105 mol dm^{-3} with respect to ethanoic acid and 0.342 mol dm^{-3} with respect to sodium ethanoate. [5]

[Total: 17]

Edexcel Specimen Unit Test 4 Q1 2000

Chapter 3

Energy changes in chemistry

The following topics are covered in this chapter:

- *Enthalpy changes*
- *Electrochemical cells*
- *Predicting redox reactions*

3.1 Enthalpy changes

LEARNING SUMMARY

After studying this section you should be able to:

- *explain and use the term 'lattice enthalpy'*
- *construct Born–Haber cycles to calculate the lattice enthalpy of simple ionic compounds*
- *explain the effect of ionic charge and ionic radius on the numerical magnitude of a lattice enthalpy*
- *calculate enthalpy changes of solution for ionic compounds from enthalpy changes of hydration and lattice enthalpies*

Key points from AS

- **Enthalpy changes**
  Revise AS pages 87–95

During the study of enthalpy changes in AS Chemistry, you learnt how to:

- calculate enthalpy changes directly from experiments using the relationship $Q = mc\Delta T$
- calculate enthalpy changes indirectly using Hess's law.

These key principles are built upon by considering the enthalpy changes that bond together an ionic lattice.

Lattice enthalpy

AQA	M5	SALTERS	M5
EDEXCEL	M4	WJEC	CH5
OCR	M5.1	NICCEA	M4
NUFFIELD	M6		

Key points from AS

- **Ionic bonding**
  Revise AS pages 45–47

Each ion is surrounded by oppositely-charged ions, forming a giant ionic lattice.

Ionic bonding is the electrostatic attraction between oppositely charged ions. This attraction acts in all directions, resulting in a giant ionic lattice with hundreds of thousands of ions (depending upon the size of the crystal). Lattice enthalpy indicates the strength of the ionic bonds in an ionic lattice.

Part of the sodium chloride lattice

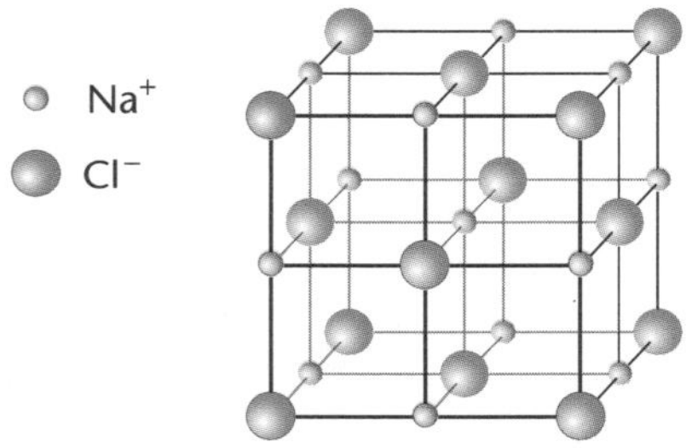

- Each Na^+ ion surrounds 6 Cl^- ions
- Each Cl^- ion surrounds 6 Na^+ ions

KEY POINT

The lattice enthalpy ($\Delta H^\ominus_{L.E.}$) of an ionic compound is the enthalpy change that accompanies the formation of 1 mole of an ionic compound from its constituent gaseous ions. ($\Delta H^\ominus_{L.E.}$ is **exothermic**.)

$$Na^+(g) + Cl^-(g) \longrightarrow Na^+Cl^-(s)$$

The term *lattice enthalpy* is also sometimes called the **enthalpy change of lattice formation.**

The opposite change, to break up the lattice, is called the **enthalpy change of lattice dissociation.**

$$Na^+Cl^-(s) \longrightarrow Na^+(g) + Cl^-(g)$$

Key points from AS

- **Indirect determination of enthalpy changes**
  Revise AS pages 91–92

Determination of lattice enthalpies

Lattice enthalpies cannot be determined directly and must be calculated indirectly using Hess's Law from other enthalpy changes that can be found experimentally. The energy cycle used to calculate a lattice enthalpy is the **Born–Haber cycle**.

The basis of the Born–Haber cycle is the formation of an ionic lattice from its elements. For sodium chloride, one route is the single stage process corresponding to its enthalpy change of formation.

ROUTE 1 $Na(s) + \frac{1}{2}Cl_2(g) \longrightarrow Na^+Cl^-(s)$

A second route can be broken down into separate steps resulting in:

ROUTE 2

- *production of gaseous atoms*
  $Na(s) \longrightarrow Na(g)$
  $\frac{1}{2}Cl_2(g) \longrightarrow Cl(g)$
- *production of gaseous ions*
  $Na(g) \longrightarrow Na^+(g) + e^-$
  $Cl(g) + e^- \longrightarrow Cl^-(g)$
- *production of the solid ionic lattice*
  $Na^+(g) + Cl^-(g) \longrightarrow Na^+Cl^-(s)$.

The Born–Haber cycle

The complete Born–Haber cycle for sodium chloride is shown below.

Definitions for these enthalpy changes are shown on p. 51.

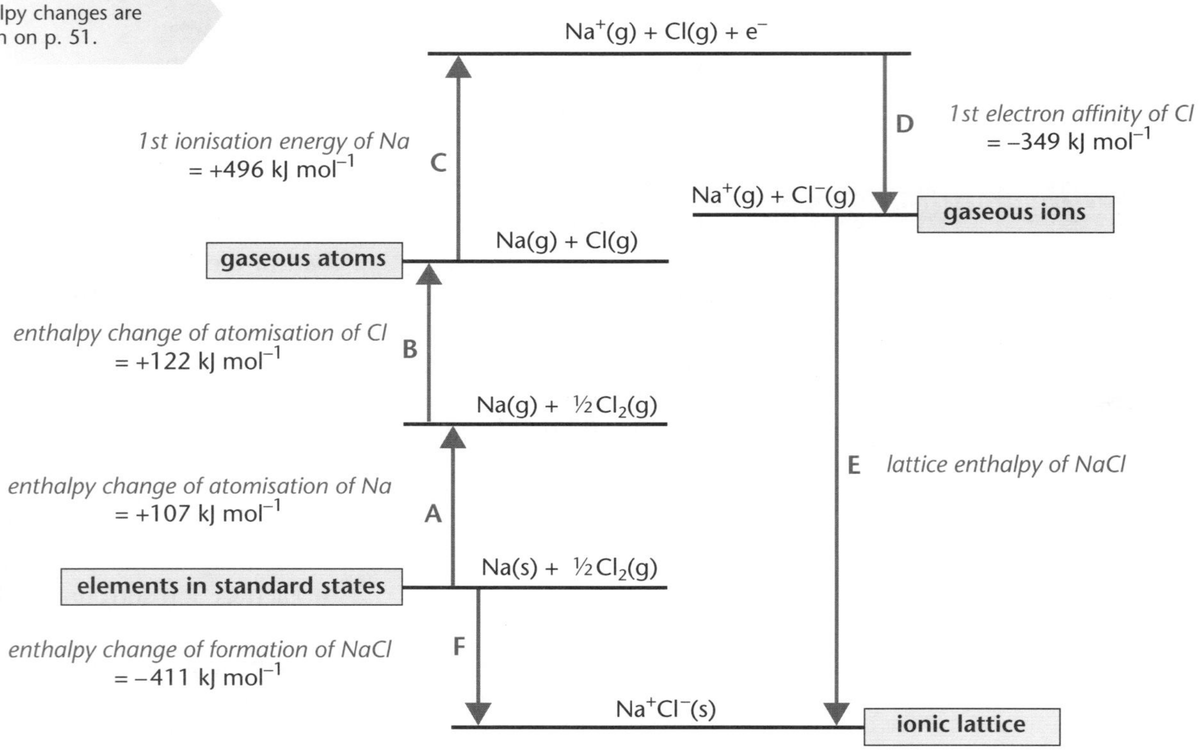

- the 1st ionisation energy of Na is **endothermic**
- the 1st electron affinity of Cl is **exothermic**.

Route 1: **A + B + C + D + E**

Route 2: **F**

Using Hess's Law: **A + B + C + D + E = F**

$\Delta H^{\ominus}_{at}\,Na(g) + \Delta H^{\ominus}_{at}\,Cl(g) + \Delta H^{\ominus}_{I.E.}\,Na(g) + \Delta H^{\ominus}_{E.A.}\,Cl(g) + \mathbf{E} = \Delta H^{\ominus}_{f}\,Na^+Cl^-(s)$

$\therefore +107 + 122 + 496 + (-349) + \mathbf{E} = -411$

Hence, the lattice energy of $Na^+Cl^-(s)$, **E** = **–787** kJ mol^{-1}

Definitions for enthalpy changes

The enthalpy changes involved in these two routes are shown below.

> **KEY POINT**
>
> **The standard enthalpy change of formation ($\Delta H^{\ominus}_{f}$)** is the enthalpy change that takes place when one mole of a compound in its standard state is formed from its constituent elements in their standard states under standard conditions.
>
> $Na(s) + \frac{1}{2}Cl_2(g) \longrightarrow Na^+Cl^-(s)$ $\qquad \Delta H^{\ominus}_{f} = -411 \text{ kJ mol}^{-1}$

Be careful when using $\Delta H^{\ominus}_{at}$ involving diatomic elements.

$\Delta H^{\ominus}_{at}$ relates to the formation of 1 mole of atoms. For chlorine, this involves $\frac{1}{2}Cl_2$ only.

> **KEY POINT**
>
> **The standard enthalpy change of atomisation ($\Delta H^{\ominus}_{at}$)** of an element is the enthalpy change that accompanies the formation of 1 mole of gaseous atoms from the element in its standard state.
>
> $Na(s) \longrightarrow Na(g)$ $\qquad \Delta H^{\ominus}_{at} = +107 \text{ kJ mol}^{-1}$
>
> $\frac{1}{2}Cl_2(g) \longrightarrow Cl(g)$ $\qquad \Delta H^{\ominus}_{at} = +122 \text{ kJ mol}^{-1}$

For gaseous molecules this enthalpy change can be determined from the **bond dissociation enthalpy** (B.D.E.) – the enthalpy change required to break and separate 1 mole of bonds so that the resulting gaseous atoms exert no forces upon each other.

$Cl\text{–}Cl(g) \longrightarrow 2Cl(g)$ $\qquad \Delta H^{\ominus}_{B.D.E.} = +244 \text{ kJ mol}^{-1}$

$\frac{1}{2}Cl\text{–}Cl(g) \longrightarrow Cl(g)$ $\qquad \Delta H^{\ominus}_{at} = +122 \text{ kJ mol}^{-1}$

> **KEY POINT**
>
> **The first ionisation energy ($\Delta H^{\ominus}_{I.E.}$)** of an element is the enthalpy change that accompanies the removal of 1 electron from each atom in 1 mole of gaseous atoms to form 1 mole of gaseous 1+ ions.
>
> $Na(g) \longrightarrow Na^+(g) + e^-$ $\qquad \Delta H^{\ominus}_{I.E} = +496 \text{ kJ mol}^{-1}$

Be careful using electron affinities. $\Delta H^{\ominus}_{E.A.}$ relates to the formation of 1 mole of 1– ions. For chlorine, this involves Cl(g) only.

> **KEY POINT**
>
> **The first electron affinity ($\Delta H^{\ominus}_{E.A.}$)** of an element is the enthalpy change that accompanies the addition of 1 electron to each atom in 1 mole of gaseous atoms to form 1 mole of gaseous 1–ions.
>
> $Cl(g) + e^- \longrightarrow Cl^-(g)$ $\qquad \Delta H^{\ominus}_{E.A.} = -349 \text{ kJ mol}^{-1}$

Key points from AS

- **Charge density**
  Revise AS page 54

Factors affecting the size of lattice enthalpies

The strength of an ionic lattice and the value of its lattice enthalpy depend upon:

- ionic size
- ionic charge.

Effect of ionic size

The effect of increasing ionic size can be seen by comparing the lattice enthalpies of sodium halides. From Cl^- to I^-, the increasing size of the negative halide ion produces a smaller charge density. This results in weaker attraction between ions and a less negative lattice enthalpy.

Lattice energy has a negative value. You should use the term '*becomes less/more negative*' instead of '*becomes bigger/smaller*' to describe any trend in lattice energy.

compound	*lattice enthalpy / kJ mol^{-1}*	*ions*	*effect of size of halide ion*
NaCl	–787	(+)(–)	ionic size increases: • charge density decreases
NaBr	–751	(+)(–)	• attraction between ions decreases
NaI	–705	(+)(–)	• lattice energy becomes less negative.

Cations are smaller than their atoms.

Anions are larger than their atoms.

Effect of ionic charge

The strongest ionic lattices with the most negative lattice enthalpies contain **small, highly charged ions.**

The diagram below shows the change in ionic size and ionic charge across Period 3 in the Periodic Table. Notice how the effects of ionic size and charge affect the size of the attraction between ions.

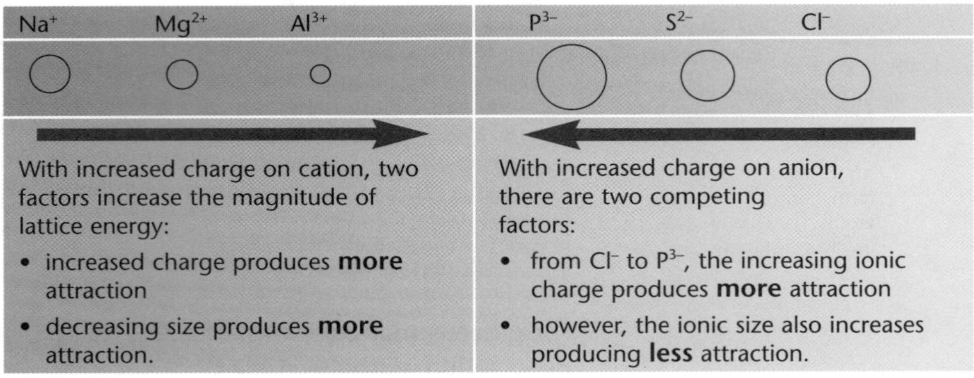
Na^+
Mg^{2+}
Al^{3+}
P^{3-}
S^{2-}
Cl^-
With increased charge on cation, two factors increase the magnitude of lattice energy:
• increased charge produces **more** attraction
• decreasing size produces **more** attraction.
With increased charge on anion, there are two competing factors:
• from Cl^- to P^{3-}, the increasing ionic charge produces **more** attraction
• however, the ionic size also increases producing **less** attraction.

Key points from AS

- **Polarisation**
  Revise AS page 54
- **The thermal stability of s-block compounds**
  Revise AS pages 74–75

Limitations of lattice enthalpies

Theoretical lattice enthalpies can be calculated by considering each ion as a perfect sphere. Any difference between this theoretical lattice enthalpy and lattice enthalpy calculated using a Born–Haber cycle indicates a degree of covalent bonding caused by polarisation. Polarisation is greatest between:

- a small densely charged cation and
- a large anion.

From the ions shown in the diagram above, we would expect the largest degree of covalency between Al^{3+} and P^{3-}.

Progress check

1 Using the information below:
(a) name each enthalpy change **A** to **E**
(b) construct a Born–Haber cycle for sodium bromide and calculate the lattice enthalpy (enthalpy change of lattice formation) of sodium bromide.

enthalpy change	*equation*	$\Delta H^{\ominus}$*/kJ mol*$^{-1}$
A	$Na(s) + \frac{1}{2}Br_2(l) \longrightarrow NaBr(s)$	–361
B	$Br(g) + e^- \longrightarrow Br^-(g)$	–325
C	$Na(s) \longrightarrow Na(g)$	+107
D	$Na(g) \longrightarrow Na^+(g) + e^-$	+496
E	$\frac{1}{2}Br_2(l) \longrightarrow Br(g)$	+112

1 (a) A enthalpy change of formation of sodium bromide
B 1st electron affinity of bromine
C enthalpy change of atomisation of sodium
D 1st ionisation energy of sodium
E enthalpy change of atomisation of bromine
(b) –751 kJ mol^{-1}

Enthalpy change of solution

AQA	M5	SALTERS	M5
EDEXCEL	M4	WJEC	CH5
NUFFIELD	M6		

An ionic lattice dissolves in **polar** solvents (e.g. water). In this process, the giant ionic lattice is broken up by polar water molecules which surround each ion in solution. The diagrams below compare the ionic lattice of sodium chloride with aqueous sodium and chloride ions.

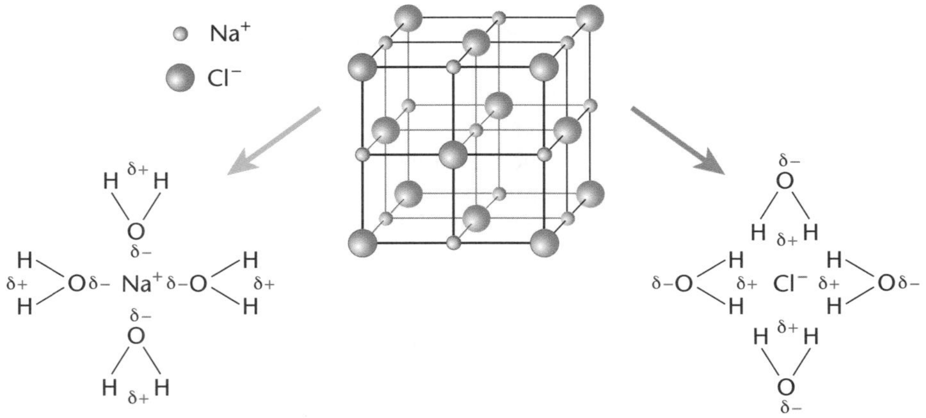

Enthalpy changes

When sodium chloride is dissolved in water, the enthalpy change of this process can be measured directly as the enthalpy change of solution.

See also complex ions p. 73.

breaking of the lattice ⟶ hydrated ions

$Na^+Cl^-(s) + aq \longrightarrow Na^+(aq) + Cl^-(aq)$ **ROUTE 1**

This change can be broken down into separate steps:

- breaking of the lattice ⟶ gaseous ions

  $Na^+Cl^-(s) \longrightarrow Na^+(g) + Cl^-(g)$
- hydration of gaseous ions ⟶ hydrated ions

  $Na^+(g) + aq \longrightarrow Na^+(aq)$

  $Cl^-(g) + aq \longrightarrow Cl^-(aq)$

} **ROUTE 2**

Key points from AS

- **Indirect determination of enthalpy changes**
  Revise AS pages 91–92

The energy cycle

The complete energy cycle for dissolving of sodium chloride in water is shown below. Definitions of these enthalpy changes are shown on page 54.

Lattice enthalpy **forms** the ionic lattice.
Notice that the energy change to break 1 mole of the ionic lattice = – (lattice enthalpy).

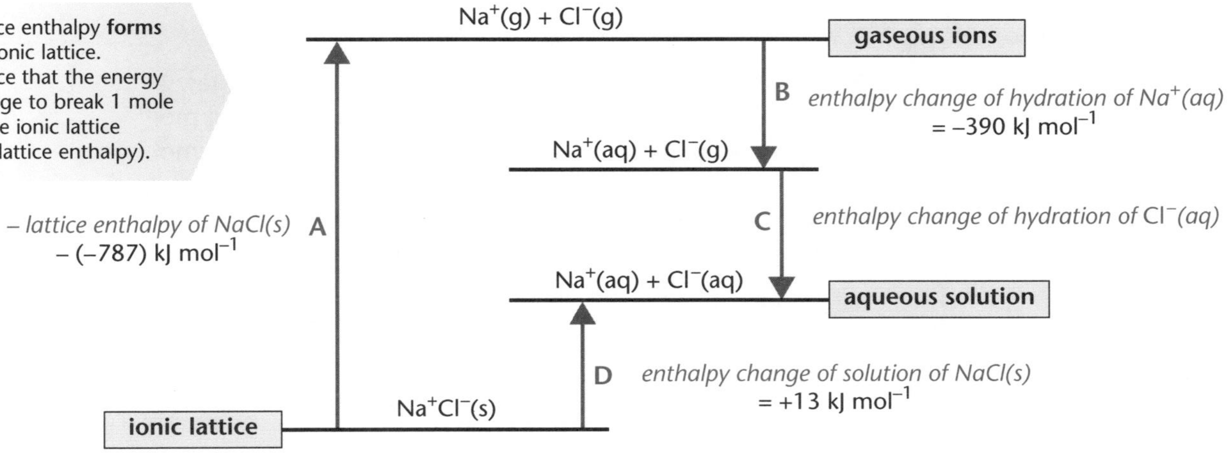

Route 1: **A + B + C**

Route 2: **D**

The production of hydrated ions comprises two changes:

- hydration of $Na^+(g)$
- hydration of $Cl^-(g)$.

Using Hess's Law: **A + B + C = D**

To determine the enthalpy change of hydration of $Cl^-(aq)$, **C**,

$$-\Delta H^{\ominus}_{L.E.}NaCl(s) + \Delta H^{\ominus}_{hyd}\ Na^+(g) + \mathbf{C} = \Delta H^{\ominus}_{solution}\ NaCl(s)$$

$$\therefore -(-787) + (-390) + \mathbf{C} = +13$$

Hence, the enthalpy change of hydration of $Cl^-(aq)$, **C = –384 kJ mol^{-1}**

Definitions for enthalpy changes

The enthalpy changes involved in these two routes are shown below.

KEY POINT

The **standard enthalpy change of solution** ($\Delta H^{\ominus}_{solution}$) is the enthalpy change that accompanies the dissolving of 1 mole of a solute in a solvent to form an infinitely dilute solution under standard conditions.

$$Na^+Cl^-(s) + aq \longrightarrow Na^+(aq) + Cl^-(aq) \qquad \Delta H^{\ominus}_{solution} = +13\ kJmol^{-1}$$

KEY POINT

The **standard enthalpy change of hydration** ($\Delta H^{\ominus}_{hyd}$) of an ion is the enthalpy change that accompanies the hydration of 1 mole of gaseous ions to form 1 mole of hydrated ions in an infinitely dilute solution under standard conditions.

$$Na^+(g) + aq \longrightarrow Na^+(aq) \qquad \Delta H^{\ominus}_{hyd} = -390\ kJmol^{-1}$$
$$Cl^-(g) + aq \longrightarrow Cl^-(aq) \qquad \Delta H^{\ominus}_{hyd} = -384\ kJmol^{-1}$$

Factors affecting the magnitude of hydration enthalpy

ion	*Na^+*	*Mg^{2+}*	*Al^{3+}*	*Cl^-*	*Br^-*	*I^-*
$\Delta H^{\ominus}_{hyd}$/kJ mol^{-1}	–390	–1891	–4613	–384	–351	–307
effect of charge	• increasing ionic charge • greater attraction for water →					
ionic radius/nm	0.102	0.072	0.053	0.180	0.195	0.215
effect of size	• decreasing ionic size • greater attraction for water →			• increasing ionic size • less attraction for water →		

The values obtained for each ion depend upon the size of the charge and the size of the ion (see also pages 51–52).

Progress check

1 You are provided with the following enthalpy changes.

$$Na^+(g) + F^-(g) \longrightarrow NaF(s) \qquad \Delta H = -918\ kJ\ mol^{-1}$$
$$Na^+(g) + aq \longrightarrow Na^+(aq) \qquad \Delta H = -390\ kJ\ mol^{-1}$$
$$F^-(g) + aq \longrightarrow F^-(aq) \qquad \Delta H = -457\ kJ\ mol^{-1}$$

(a) Name each of these three enthalpy changes.

(b) Write an equation, including state symbols, for which the enthalpy change is the enthalpy change of solution of NaF.

(c) Calculate the enthalpy change of solution of NaF.

1 (a) lattice enthalpy of NaF
enthalpy change of hydration of Na^+
enthalpy change of hydration of F^-
(b) $NaF(s) + aq \longrightarrow Na^+(aq) + F^-(aq)$
(c) +71 kJ mol^{-1}

3.2 Electrochemical cells

After studying this section you should be able to:

- *understand the principles of an electrochemical cell*
- *define the term 'standard electrode (redox) potential', $E^{\ominus}$*
- *describe how to measure standard electrode potentials*
- *calculate a standard cell potential from standard electrode potentials*

LEARNING SUMMARY

Key points from AS

- **Redox reactions**
  Revise AS pages 68–70

During the study of redox in AS Chemistry you learnt:

- that oxidation and reduction involve transfer of electrons in a redox reaction
- the rules for assigning oxidation states
- how to construct an overall redox equation from two half-equations.

These key principles are built upon by considering electrochemical cells in which redox reactions provide electrical energy.

Electricity from chemical reactions

AQA	M5	SALTERS	M4
EDEXCEL	M5	WJEC	CH5
OCR	M5.6	NICCEA	M4
NUFFIELD	M6		

In a redox reaction, electrons are transferred between the reacting chemicals.

For example, zinc react with aqueous copper(II) ions in a redox reaction.

$$Zn(s) + Cu^{2+}(aq) \longrightarrow Zn^{2+}(aq) + Cu(s)$$

oxidation: $Zn(s) \longrightarrow Zn^{2+}(aq) + 2e^-$
reduction: $Cu^{2+}(aq) + 2e^- \longrightarrow Cu(s)$

Electrochemical cells are the basis of all batteries. Their true value is as a portable form of electricity.

If this reaction is carried out simply by mixing zinc with aqueous copper(II) ions:

- heat energy is produced and, unless used as useful heat, this is often wasted.

An electrochemical cell controls the transfer of electrons:

- electrical energy is produced and can be used for electricity.

The zinc-copper electrochemical cell

An electrochemical cell comprises two **half-cells**. In each half-cell, the **oxidation** and **reduction** processes take place **separately**.

A half-cell contains the oxidised and reduced species from a half-equation.
The species can be converted into one another by addition or removal of electrons.

A zinc-copper cell has two half-cells:

- an oxidation half-cell containing reactants and products from the oxidation half-reaction: $Zn^{2+}(aq)$ and $Zn(s)$
- a reduction half-cell containing reactants and products from the reduction half-reaction: $Cu^{2+}(aq)$ and $Cu(s)$.

The circuit is completed using:

- a connecting **wire** which transfers **electrons** between the half-cells
- a **salt bridge** which transfers **ions** between the half-cells.

The salt bridge consists of filter paper soaked in aqueous KNO_3 or NH_4NO_3.

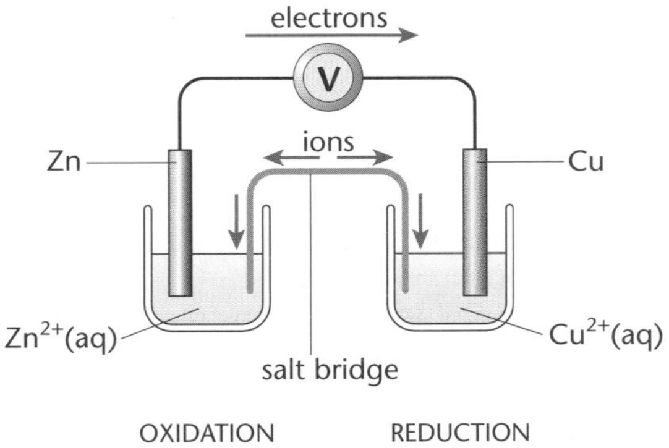

Remember these charge carriers:
ELECTRONS through the wire;
IONS through the salt bridge.

oxidation half-cell: $Zn^{2+}(aq)$ and Zn(s)

- **Zn is oxidised** to Zn^{2+}: $Zn(s) \longrightarrow Zn^{2+}(aq) + 2e^-$
- electrons are **supplied** to the external circuit
- the polarity is **negative**.

reduction half-cell: $Cu^{2+}(aq)$ and Cu(s)

- **Cu^{2+} is reduced** to Cu: $Cu^{2+}(aq) + 2e^- \longrightarrow Cu(s)$
- electrons are **taken** from the external circuit
- the polarity is **positive**.

The measured cell e.m.f. of 1.10 V indicates that there is a potential difference of 1.10 V between the copper and zinc half-cells.

Standard electrode potentials

AQA	M5	SALTERS	M4
EDEXCEL	M5	WJEC	CH5
OCR	M5.6	NICCEA	M4
NUFFIELD	M6		

A **hydrogen half-cell** is used as the standard for the measurement of standard electrode potentials.

The standard hydrogen half-cell is based upon the half-reaction below.

$$H^+(aq) + e^- \rightleftharpoons \tfrac{1}{2}H_2(g)$$

A hydrogen half-cell typically comprises

- 1 mol dm^{-3} hydrochloric acid as the source of $H^+(aq)$
- a supply of hydrogen gas, $H_2(g)$ at 100 kPa
- an inert platinum electrode on which the half-reaction above takes place.

KEY POINT

The **standard electrode potential** of a half-cell, $E^\ominus$, is the e.m.f. of a half-cell compared with a standard hydrogen half-cell.
All measurements are at **298 K** with solution **concentrations of 1 mol dm^{-3}** and gas **pressures of 100 kPa.**

Measuring standard electrode potentials

Half-cells comprising a metal and a metal ion

The diagram below shows how a hydrogen half-cell can be used to measure the standard electrode potential of a copper half-cell.

A high-resistance voltmeter is used to minimise the current that flows.

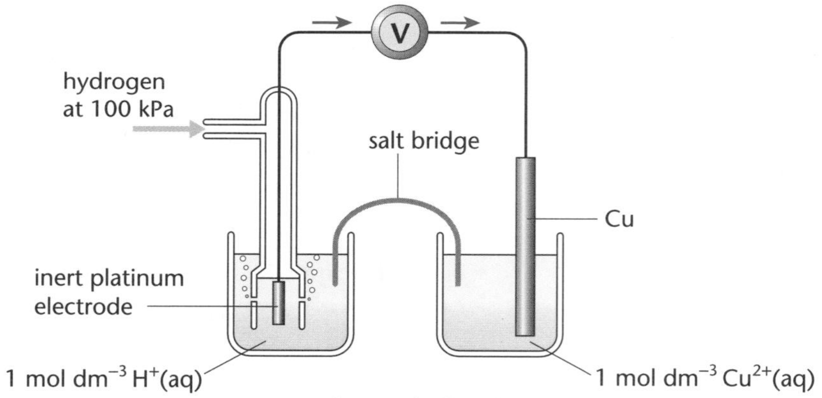

Hydrogen half-cell

KEY POINT

- The standard electrode potential of a half-cell represents its contribution to the cell e.m.f.
- The contribution made by the hydrogen half-cell to the cell e.m.f. is defined as 0 V.
- The **sign** of the standard electrode potential of a half-cell indicates its **polarity** compared with the hydrogen half-cell.

As with the hydrogen or platinum electrode half-cell, provides a surface on which the half-reaction can take place.

Half-cells comprising ions of different oxidation states

The half-reaction taking place in a half-cell can be between aqueous ions with different oxidation states. To allow electrons to pass into the half-cell, an inert electrode of platinum is used.

For example, a half-cell containing $Fe^{3+}(aq)$ and $Fe^{2+}(aq)$ ions is based upon the half-reaction below.

$$Fe^{3+}(aq) \rightleftharpoons Fe^{2+}(aq) + e^-$$

A **standard** $Fe^{3+}(aq)/Fe^{2+}(aq)$ half-cell requires:

- a solution containing **both**
  - 1 mol dm^{-3} $Fe^{3+}(aq)$ **and**
  - 1 mol dm^{-3} $Fe^{2+}(aq)$
- an inert platinum electrode with a connecting wire.

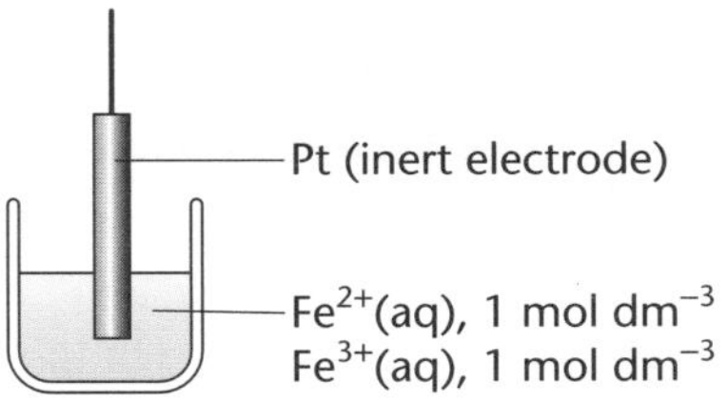

The electrochemical series

AQA	M5	SALTERS	M4
EDEXCEL	M5	WJEC	CH5
OCR	M5.6	NICCEA	M4
NUFFIELD	M6		

A standard electrode potential indicates the availability of electrons from a half-reaction.

Metals are reducing agents.

- Metals typically react by donating electrons in a redox reaction.
- Reactive metals have the greatest tendency to donate electrons and their half-cells have the most negative electrode potentials.

Non-metals are oxidising agents.

- Non-metals typically react by accepting electrons in a redox reaction.
- Reactive non-metals have the greatest tendency to accept electrons and their half-cells have the most positive electrode potentials.

Key points from AS

- **Reactivity of s-block elements**
  Revise AS page 72
- **The relative reactivity of the halogens as oxidising agents**
  Revise AS page 77

The **electrochemical series** shows electrode reactions listed in order of their standard electrode potentials. The electrochemical series below shows standard electrode potentials listed with the **most positive** $E^⊖$ value at the top. This enables oxidising and reducing agents to be easily compared.

This version of the electrochemical series show standard electrode potentials listed with the **most positive** $E^⊖$ value at the top.

You will find versions of the electrochemical series listed with the **most negative** $E^⊖$ value at the top.

This does not matter. The important point is that the potentials are listed **in order** and **can be compared**.

	electrode reaction					$E^⊖$/ V	
strongest oxidising agent ↑	$F_2(g)$	+	$2e^-$	$\rightleftharpoons$	$2F^-(aq)$	+2.87	↓
	$Cl_2(g)$	+	$2e^-$	$\rightleftharpoons$	$2Cl^-(aq)$	+1.36	
	$Br_2(l)$	+	$2e^-$	$\rightleftharpoons$	$2Br^-(aq)$	+1.07	
	$Ag^+(aq)$	+	e^-	$\rightleftharpoons$	$Ag(s)$	+0.80	
	$Cu^{2+}(aq)$	+	$2e^-$	$\rightleftharpoons$	$Cu(s)$	+0.34	
	$H^+(aq)$	+	e^-	$\rightleftharpoons$	$\frac{1}{2}H_2(g)$	0	
	$Fe^{2+}(aq)$	+	$2e^-$	$\rightleftharpoons$	$Fe(s)$	−0.44	
	$Cr^{3+}(aq)$	+	$3e^-$	$\rightleftharpoons$	$Cr(s)$	−0.77	*strongest reducing agent*
	$K^+(aq)$	+	e^-	$\rightleftharpoons$	$K(s)$	−2.92	

KEY POINT

In the electrochemical series:
- the **oxidised** form is on the **left-hand side** of the half-equation
- the **reduced** form is on the **right-hand side** of the half-equation.

Standard cell potentials and cell reactions

AQA	M5	SALTERS	M4
EDEXCEL	M5	WJEC	CH5
OCR	M5.6	NICCEA	M4
NUFFIELD	M6		

Key points from AS

- **Combining half-equations**
  Revise AS page 70

The **standard cell potential** of a cell is the e.m.f. acting between the two half-cells making up the cell under standard conditions. A standard cell potential can be determined using standard electrode potentials.

The **cell reaction** is:

- the overall process taking place in the cell
- the sum of the reduction and oxidation half-reactions taking place in each half-cell.

Calculating the standard cell potential of a silver-iron(II) cell

Identify the two relevant half-reactions and the polarity of each electrode.

The more positive of the two systems is the positive terminal of the cell.

$Ag^{+}(aq) + e^{-} \rightleftharpoons Ag(s) \quad E^{\ominus} = +0.80\ V$ *positive terminal*

$Fe^{2+}(aq) + 2e^{-} \rightleftharpoons Fe(s) \quad E^{\ominus} = -0.44\ V$ *negative terminal*

The standard electrode potential is the difference between the $E^{\ominus}$ values:

$E^{\ominus}_{cell}$ is the difference between the standard electrode potentials.

Simply subtract the $E^{\ominus}$ of the negative terminal from the $E^{\ominus}$ of the positive terminal:

$$E^{\ominus}_{cell} = E^{\ominus}\ (\text{positive terminal}) - E^{\ominus}\ (\text{negative terminal})$$

$$\therefore\ E^{\ominus}_{cell} = 0.080 - (-0.44) = \mathbf{1.24\ V}$$

Determination of the cell reaction in a silver-iron(II) cell

The more negative half-equation is reversed. The negative terminal is then on the left-hand side.

Work out the actual direction of the half-equations.

The **half-equation** with the **more negative $E^{\ominus}$** value provides the electrons and proceeds to the left. This half-equation is reversed so that the two half-reactions taking place can be clearly seen and compared.

$Ag^{+}(aq) + e^{-} \longrightarrow Ag(s)$ *reaction at positive terminal*

$Fe(s) \longrightarrow Fe(aq) + 2e^{-}$ *reaction at negative terminal*

This is discussed in detail in the AS Chemistry Study Guide, p. 70.

The overall cell reaction can be found by adding the half-equations.

The silver half-equation must first be multiplied by '2' to balance the electrons.

The two half-equations are then added.

$$2Ag^{+}(aq) + Fe(s) \longrightarrow 2Ag(s) + Fe^{2+}(aq)$$

Progress check

1 Use the standard electrode potentials on page 57, to answer the questions which follow.

(a) Calculate the standard cell potential for the following cells.

(i) F_2/F^- and Ag^+/Ag

(ii) Ag^+/Ag and Cr^{3+}/Cr

(iii) Cu^{2+}/Cu and Cr^{3+}/Cr.

(b) For each cell in (a) determine which half-reactions proceed and hence construct the cell reaction.

1 (a) (i) 2.07 V (ii) 1.57 V (iii) 1.11 V.

(b) $F_2 + 2Ag \longrightarrow 2F^- + 2Ag^+$

$3Ag^+ + Cr \longrightarrow 3Ag + Cr^{3+}$

$3Cu^{2+} + 2Cr \longrightarrow 3Cu + 2Cr^{3+}$

3.3 Predicting redox reactions

LEARNING SUMMARY

After studying this section you should be able to:

- *use standard cell potentials to predict the feasibility of a reaction*
- *recognise the limitations of such predictions*

During AS Chemistry, you learnt how to combine two half-equations to construct a full redox equation.

Standard electrode potentials can be used to predict the direction of any aqueous redox reaction from the relevant half-reactions.

Using standard electrode potentials to make predictions

AQA	M5	SALTERS	M4
EDEXCEL	M5	WJEC	CH5
OCR	M5.6	NICCEA	M4
NUFFIELD	M6		

Key points from AS

- **Redox reactions**
  Revise AS pages 68–70

By studying standard electrode potentials, predictions can be made about the likely feasibility of a reaction.

The table below shows three redox systems.

	redox system	$E^\ominus$ /V
A	$H_2O_2(aq) + 2H^+ + 2e^- \rightleftharpoons 2H_2O(l)$	+1.77
B	$I_2(aq) + 2e^- \rightleftharpoons 2I^-(aq)$	+0.54
C	$Fe^{3+}(aq) + 3e^- \rightleftharpoons Fe(s)$	–0.04

We can predict the redox reactions that may take place between these species by treating the redox systems as if they are half-cells.

Notice that a reaction takes place between reactants from different sides of each half equation.

Comparing the redox systems A and B

- The I_2/I^- system has the less positive $E^\ominus$ value and its half-reaction will proceed to the left, donating electrons:

$$\mathbf{H_2O_2}(aq) + \mathbf{2H^+}(aq) + 2e^- \longrightarrow 2H_2O(l) \qquad E^\ominus = +1.77\text{ V}$$
$$I_2(aq) + 2e^- \longleftarrow 2\mathbf{I^-}(aq) \qquad E^\ominus = +0.54\text{ V}$$

We would therefore predict that $\mathbf{H_2O_2}$ **and** $\mathbf{H^+}$ would react with $\mathbf{I^-}$.

Comparing the redox systems B and C

- The Fe^{3+}/Fe system has the more negative $E^\ominus$ value and its half-reaction will proceed to the left, donating electrons:

$$\mathbf{I_2}(aq) + 2e^- \longrightarrow 2I^-(aq) \qquad E^\ominus = +0.54\text{ V}$$
$$Fe^{3+}(aq) + 3e^- \longleftarrow \mathbf{Fe}(s) \qquad E^\ominus = -0.04\text{ V}$$

We would therefore predict that $\mathbf{I_2}$ would react with **Fe**.

Comparing the redox systems A and C

- The Fe^{3+}/Fe system has the more negative $E^\ominus$ value and its half-reaction will proceed to the left, donating electrons:

$$\mathbf{H_2O_2}(aq) + \mathbf{2H^+}(aq) + 2e^- \longrightarrow 2H_2O(l) \qquad E^\ominus = +1.77\text{ V}$$
$$Fe^{3+}(aq) + 3e^- \longleftarrow \mathbf{Fe}(s) \qquad E^\ominus = -0.04\text{ V}$$

We would therefore predict that $\mathbf{H_2O_2}$ **and** $\mathbf{H^+}$ would react with **Fe**.

KEY POINT

When comparing two redox systems, we can predict that a reaction may take place between:
- the stronger reducing agent with the more negative $E^\ominus$, on the right-hand side of the redox system
- the stronger oxidising agent with the more positive $E^\ominus$, on the left-hand side of the redox system .

Limitations of standard electrode potentials

AQA	M5	SALTERS	M4
EDEXCEL	M5	WJEC	CH5
OCR	M5.6	NICCEA	M4
NUFFIELD	M6		

Electrode potentials and ionic concentration

Non-standard conditions alter the value of an electrode potential.

The half-equation for the copper half-cell is shown below.

$$Cu^{2+}(aq) + 2e^- \rightleftharpoons Cu(s)$$

Using le Chatelier's principle, by increasing the concentration of $Cu^{2+}(aq)$:

- the equilibrium opposes the change by moving to the right
- electrons are removed from the equilibrium system
- the electrode potential becomes more positive.

A change in electrode potential resulting from concentration changes means that predictions made on the basis of the **standard** value may not be valid.

Will a reaction actually take place?

Remember that these are equilibrium processes.

- Predictions can be made but these give no indication of the reaction rate which may be extremely slow, caused by a large activation energy.
- The actual conditions used may be different from the standard conditions used to record $E^{\ominus}$ values. This will affect the value of the electrode potential (see above).
- Standard electrode potentials apply to aqueous equilibria. Many reactions take place that are not aqueous.

As a general working rule:

- the larger the difference between $E^{\ominus}$ values, the more likely that a reaction will take place
- if the difference between $E^{\ominus}$ values is less than 0.4 V, then a reaction is unlikely to take place.

Progress check

1 For each of the three redox systems on page 59, construct full redox equations.

2 Use the standard electrode potentials below to answer the questions which follow.

$$Fe^{3+}(aq) + e^- \rightleftharpoons Fe^{2+}(aq) \qquad E^{\ominus} = +0.77\ V$$
$$Br_2(l) + 2e^- \rightleftharpoons 2Br^-(aq) \qquad E^{\ominus} = +1.07\ V$$
$$Ni^{2+}(aq) + 2e^- \rightleftharpoons Ni(s) \qquad E^{\ominus} = -0.25\ V$$
$$O_2(g) + 4H^+(aq) + 4e^- \rightleftharpoons 2H_2O(l) \qquad E^{\ominus} = +1.23\ V$$

(a) Arrange the reducing agents in order with the most powerful first.

(b) Predict the reactions that could take place.

1 **A** and **B**: $H_2O_2(aq) + 2H^+(aq) + 2I^-(aq) \longrightarrow I_2(aq) + 2H_2O(l)$
B and **C**: $2Fe(s) + 3I_2(aq) \longrightarrow 2Fe^{3+}(aq) + 6I^-(aq)$
A and **C**: $3H_2O_2(aq) + 6H^+(aq) + 2Fe(s) \longrightarrow 2Fe^{3+}(aq) + 6H_2O(l)$

2 $Ni(s)$, $Fe^{2+}(aq)$, $Br^-(l)$, $H_2O(l)$.
Br_2 with Fe^{2+}; Fe^{3+} with Ni; O_2 and H^+ with Fe^{2+}; Br_2 with Ni;
O_2 and H^+ with Br^-; O_2 and H^+ with Ni.

Sample question and model answer

Take care to include correct state symbols – essential when showing the equations representing these enthalpy changes.

(a) The lattice energy of copper(II) oxide can be found using a Born–Haber cycle. Write equations, including state symbols, which correspond to:

(i) the enthalpy change of formation of copper(II) oxide

$Cu(s) + \frac{1}{2}O_2(g) \longrightarrow CuO(s)$ formulae ✓ state symbols ✓

(ii) the lattice energy of copper(II) oxide

$Cu^{2+}(g) + O^{2-}(g) \longrightarrow CuO(s)$ formulae ✓ state symbols ✓

(iii) the first ionisation energy of copper.

$Cu(g) \longrightarrow Cu^{+}(g) + e^{-}$ formulae ✓ state symbols ✓ [6]

(b) Use the data below to construct a Born–Haber cycle and hence calculate a value for the lattice energy of copper(II) oxide.

enthalpy change	***/ kJ mol^{-1}***
atomisation of copper	+339
1st ionisation energy of copper	+745
2nd ionisation energy of copper	+1960
atomisation of oxygen, i.e. $\frac{1}{2}O_2(g) \rightarrow O(g)$	+248
1st electron affinity of oxygen	–141
2nd electron affinity of oxygen	+791
formation of copper(II) oxide	–155

copper part of cycle ✓
oxygen part of cycle ✓
CuO part of cycle ✓

This Born–Haber cycle includes all the enthalpy changes that you may encounter in a Born–Haber cycle.

Most cycles in exams are simpler.

Note the different signs for the 1st and 2nd electron affinities.

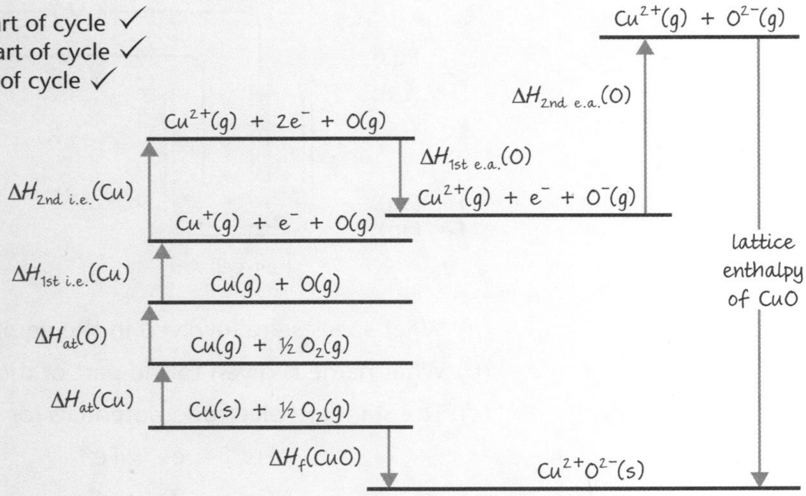

By Hess's Law.

$\Delta H_{at}(Cu) + \Delta H_{at}(O) + \Delta H_{1st\ i.e.}(Cu) + \Delta H_{2nd\ i.e.}(Cu) + \Delta H_{1st\ e.a.}(O) + \Delta H_{2nd\ e.a.}(O) + L.E.(CuO) = \Delta H_f(CuO)$

$\therefore (+339) + (+248) + (+745) + (+1960) + (-141) + (+791) + L.E.(CuO) = (-155)$

$\therefore$ Lattice energy CuO = -4097 kJ mol^{-1} correct value correct sign ✓✓

[Total: 11]

Cambridge How Far, How Fast? Q7 Modified June 1998

Practice examination questions

1

(a) Construct a Born–Haber cycle for the formation of sodium oxide, Na_2O, and use the data given below to calculate the second electron affinity of oxygen.

Na(s)	$\longrightarrow$	Na(g)	$\Delta H^\ominus = +107\ kJ\ mol^{-1}$
Na(g)	$\longrightarrow$	$Na^+(g) + e^-$	$\Delta H^\ominus = +496\ kJ\ mol^{-1}$
$O_2(g)$	$\longrightarrow$	2O(g)	$\Delta H^\ominus = +249\ kJ\ mol^{-1}$
$O(g) + e^-$	$\longrightarrow$	$O^-(g)$	$\Delta H^\ominus = -141\ kJ\ mol^{-1}$
$O^-(g) + e^-$	$\longrightarrow$	$O^{2-}(g)$	$\Delta H^\ominus$ to be calculated
$2Na^+(g) + O^{2-}(g)$	$\longrightarrow$	$Na_2O(s)$	$\Delta H^\ominus = -2478\ kJ\ mol^{-1}$
$2Na(s) + \frac{1}{2}O_2(g)$	$\longrightarrow$	$Na_2O(s)$	$\Delta H^\ominus = -414\ kJ\ mol^{-1}$ [9]

(b) Explain why the second electron affinity of oxygen has a large positive value. [2]

(c) Explain, by reference to steps from relevant Born–Haber cycles, why sodium forms a stable oxide consisting of Na^+ and O^{2-} ions but not oxides consisting of Na^+ and O^- or Na^{2+} and O^{2-} ions. [4]

[Total: 15]

Assessment and Qualifications Alliance Further Physical Chemistry Q2 June 2000

2

A cell was set up as shown below.

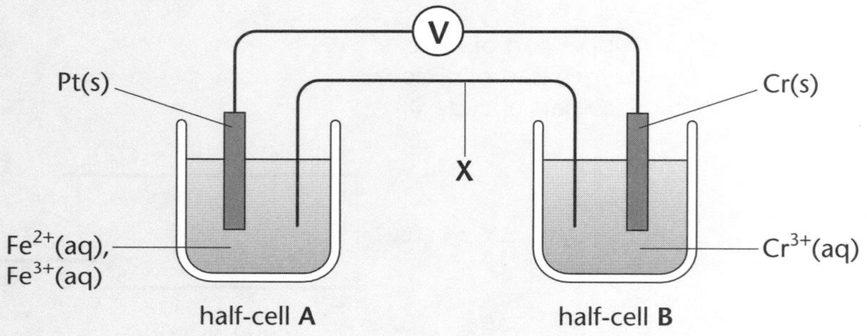

(a) What species are involved in the equilibria in the separate half-cells? [2]

(b) What name is given to the part of the cell labelled X? [1]

(c) The standard electrode potentials for the half-cells above are:

$Fe^{3+} + e^- \rightleftharpoons Fe^{2+}$ $\quad E^\ominus = +0.77\ V$

$Cr^{3+} + 3e^- \rightleftharpoons Cr$ $\quad E^\ominus = -0.74\ V$

(i) Calculate the standard cell potential of the complete cell.

(ii) Draw an arrow on the diagram above to show the direction of electron flow in the external circuit. [2]

(d) (i) Write equations for the reaction at each electrode. [2]

(ii) Hence write an equation to show the overall reaction. [1]

[Total: 8]

Cambridge How Far, How Fast? Q2 March 1998

Practice examination questions (continued)

3

(a) Use the standard electrode potentials given below to answer the questions which follow.

			$E^\ominus$/V
$Fe^{3+}(aq) + e^-$	$\rightleftharpoons$	$Fe^{2+}(aq)$	+0.77
$I_2(aq) + 2e^-$	$\rightleftharpoons$	$2I^-(aq)$	+0.54
$[Fe(CN)_6]^{3-}(aq) + e^-$	$\rightleftharpoons$	$[Fe(CN)_6]^{4-}(aq)$	+0.34

(i) Which of the ions $Fe^{3+}(aq)$ and $[Fe(CN)_6]^{3-}(aq)$ is the stronger oxidising agent?

(ii) State, with an equation and a reason, which of the iron-containing species shown above would be capable of liberating iodine when added to an aqueous solution containing iodide ions. [4]

(b) Consider the following standard electrode potentials.

					$E^\ominus$/V
$Sn^{2+}(aq)$	+	$2e^-$	$\rightleftharpoons$	$Sn(s)$	–0.14
$Fe^{2+}(aq)$	+	$2e^-$	$\rightleftharpoons$	$Fe(s)$	–0.44
$Zn^{2+}(aq)$	+	$2e^-$	$\rightleftharpoons$	$Zn(s)$	–0.76

Some cars are made from coated steel. Use the above data to explain why zinc, rather than tin, is used for this coating. [2]

[Total: 6]

NEAB Paper 1B Q5 June 1993 modified

4

(a) Name the standard reference electrode against which electrode potentials are measured and, for this electrode, state the conditions to which the term standard refers. [4]

(b) The standard electrode potentials for two electrode reactions are given below.

$S_2O_8^{2-}(aq) + 2e^- \longrightarrow 2SO_4^{2-}(aq)$ $\quad E^\ominus = +2.01$ V

$Ag^+(aq) + e^- \longrightarrow Ag(s)$ $\quad E^\ominus = +0.80$ V

(i) A cell is produced when these two half-cells are connected. Deduce the cell potential, $E^\ominus$, for this cell and write an equation for the spontaneous reaction.

(ii) State how, if at all, the electrode potential of the $S_2O_8^{2-}/SO_4^{2-}$ equilibrium would change if the concentration of SO_4^{2-} ions was increased. Explain your answer. [6]

[Total: 10]

Assessment and Qualifications Alliance Further Physical Chemistry Q1 March 2000

Chapter 4

The Periodic Table

The following topics are covered in this chapter:

- *Oxides of the Period 3 elements*
- *Chlorides of the Period 3 elements*
- *The transition elements*
- *Transition element complexes*
- *Ligand substitution of complex ions*
- *Redox reactions of transition metal ions*
- *Catalysis*
- *Reactions of metal aqua-ions*

Key points from AS

- **Bonding and structure**
  Revise AS pages 44–61

During AS Chemistry, you learnt about physical trends in properties of **elements** across Period 3. Across Period 3, **compounds** of the elements also show characteristic trends. For A2 Chemistry, physical and chemical trends in the properties of Period 3 oxides and chlorides are explained in terms of bonding and structure.

4.1 Oxides of the Period 3 elements

After studying this section you should be able to:

- *describe and explain the periodic variation of the formulae and boiling points of the main oxides of Period 3*
- *describe the reactions of Period 3 elements with oxygen*
- *describe the action of water on Period 3 oxides*
- *describe these reactions in terms of structure and bonding*

LEARNING SUMMARY

The formulae of the Period 3 oxides

AQA	M5	SALTERS	M5
EDEXCEL	M4	WJEC	CH5
OCR	M5.1	NICCEA	M4
NUFFIELD	M6		

Key points from AS

- **The modern Periodic Table**
  Revise AS pages 64–67
- **Redox reactions**
  Revise AS pages 68–70

The highest oxidation state of an element is usually the group number.

Across Period 3, the formulae of compounds show a regular pattern which depends upon the **number of electrons** in the outer shell.

The **oxidation state** of an element indicates the **number of electrons** involved in **bonding**. The relationship of the formulae of the Period 3 oxides with electronic structure and oxidation state is shown below.

element	Na	Mg	Al	Si	P	S
electrons in outer shell	1	2	3	4	5	6
formula of oxide	Na_2O	MgO	Al_2O_3	SiO_2	P_4O_6 P_4O_{10}	SO_2 SO_3
oxidation state of element in oxide	+1	+2	+3	+4	+3 +5	+4 +6

increase in highest oxidation state →

The amount of oxygen atoms per mole of the element increases steadily across Period 3.

For each element in the Period, there is an increase of 0.5 mole of O per mole of element.

$Na_1O_{0.5}$ → Mg_1O_1 → $Al_1Cl_{1.5}$ → Si_1O_2 → $P_1O_{2.5}$ → S_1O_3

Periodicity means a repeating trend. This is one of the important ideas of chemistry – a trend across one period is repeated across other periods.

This periodicity is also seen across Period 2 and 4:

Period 2	Li_2O	BeO	Be_2O_3	CO_2
Period 3	Na_2O	MgO	Al_2O_3	SiO_2
Period 4	K_2O	CaO	Ga_2O_3	GeO_2

Preparation of oxides from the elements

AQA	M5	SALTERS	M5
EDEXCEL	M4	WJEC	CH5
OCR	M5.1	NICCEA	M4
NUFFIELD	M6		

Oxides of the Period 3 elements Na to Cl can be prepared by heating each element in oxygen. These are redox reactions:

- the Period 3 element is oxidised
- oxygen is reduced.

Metal oxides

Equations for the reactions of oxygen with the metals sodium, magnesium and aluminium are shown below.

$$4Na(s) + O_2(g) \xrightarrow{\text{yellow flame}} 2Na_2O(s)$$

$$2Mg(s) + O_2(g) \xrightarrow{\text{white flame}} 2MgO(s)$$

$$4Al(s) + 3O_2(g) \xrightarrow{\text{white flame}} 2Al_2O_3(s)$$

- Na_2O and MgO are **ionic compounds**.
- Al_2O_3 has bonding **intermediate** between ionic and covalent.

At room temperature and pressure:
Na_2O, MgO and Al_2O_3 are white solids.

Phosphorus and sulphur exist as P_4 and S_8 molecules.
These are usually simplified in equations as P and S.

Non-metal oxides

Equations for the reactions of phosphorus and sulphur with oxygen are shown below.

$$4P(s) + 5O_2(g) \xrightarrow{\text{white flame}} P_4O_{10}(s)$$

$$S(s) + O_2(g) \xrightarrow{\text{blue flame}} SO_2(g)$$

SO_2 reacts further with oxygen in the presence of a catalyst (V_2O_5) see page 83:

$$2SO_2(g) + O_2(g) \longrightarrow 2SO_3(l)$$

- P_4O_{10}, SO_2 and SO_3 are **covalent compounds**.

At room temperature and pressure:
- P_4O_{10} is a white solid
- SO_2 is a colourless gas
- SO_3 is a colourless liquid.

Trend in the boiling points of the Period 3 oxides

AQA	M5	SALTERS	M5
EDEXCEL	M4	WJEC	CH5
OCR	M5.1	NICCEA	M4
NUFFIELD	M6		

Key point from AS

- **Bonding, structure and properties**
  Revise AS pages 58–61
- **Trends in boiling points of Period 3 elements**
  Revise AS pages 66–67

The table below compares the boiling points of some Period 3 oxides with their structure and bonding.

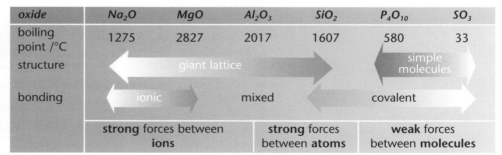

oxide	*Na_2O*	*MgO*	*Al_2O_3*	*SiO_2*	*P_4O_{10}*	*SO_3*
boiling point /°C	1275	2827	2017	1607	580	33
structure	giant lattice				simple molecules	
bonding	ionic		mixed	covalent		
	strong forces between **ions**		**strong** forces between **atoms**		**weak** forces between **molecules**	

Giant structures and boiling point

The oxides with **giant** structures ($Na_2O \rightarrow SiO_2$):

- have **high boiling points**
- require a large amount of energy to break the **strong forces** between their particles.

The particles making up a giant lattice can be different.

- Na_2O and MgO have ionic bonding and the **strong** forces are electrostatic attractions acting between positive and negative **ions**.
- Al_2O_3 has bonding intermediate between ionic and covalent. **Strong** forces act between particles **intermediate** between ions and atoms.
- SiO_2 has covalent bonding and the **strong** forces are 'shared pairs of electrons' acting between the **atoms**.

During your study of AS Chemistry, you learnt that the boiling points of the elements and compounds are dependent upon bonding and structure.
The same principles can be applied many times in chemistry covalent bonds.

Simple molecular structures and boiling point

The oxides with **simple molecular** structures ($P_4O_{10} \rightarrow SO_3$):

- have low boiling points
- require a small amount of energy to break the **weak van der Waals' forces** acting between the **molecules**.

The action of water on the Period 3 oxides

AQA	M5	SALTERS	M5
EDEXCEL	M4	WJEC	CH5
OCR	M5.1	NICCEA	M4
NUFFIELD	M6		

Reactions of sodium and magnesium oxides

Metal oxides form alkalis in water.

$$Na_2O(s) + 2H_2O(l) \longrightarrow 2NaOH(aq) \qquad pH = 14$$

$$MgO(s) + 2H_2O(l) \longrightarrow Mg(OH)_2(aq) \qquad pH = 11$$

Each hydroxide dissociates in water, releasing OH^- ions into solution:

$$NaOH(aq) \longrightarrow Na^+(aq) + OH^-(aq)$$

Although MgO reacts with water, the magnesium hydroxide formed is only slightly soluble in water forming a weak alkaline solution only.

Oxides of aluminium and silicon

Al_2O_3 and SiO_2 have very strong lattices which cannot be broken down by water. Consequently these compounds are **insoluble** in water.

Key points from AS

- **The Group 1 elements** *Revise AS pages 72–73*
- **The Group 2 elements** *Revise AS pages 73–74*

Reactions of non-metal oxides

Non-metal oxides form acids in water. Equations for the reactions of $P_4O_{10}(s)$, $SO_2(g)$ and $SO_3(l)$ with water are shown below.

$$P_4O_{10}(s) + 6H_2O(l) \longrightarrow 4H_3PO_4(aq) \qquad \textit{phosphoric acid}$$

$$SO_2(g) + H_2O(l) \longrightarrow H_2SO_3(aq) \qquad \textit{sulphurous acid}$$

$$SO_3(l) + H_2O(l) \longrightarrow H_2SO_4(aq) \qquad \textit{sulphuric acid}$$

Each acid dissociates in the water, releasing H^+ ions into solution.

E.g. $H_2SO_4(aq) \longrightarrow H^+(aq) + HSO_4^-(aq)$

Metal oxides are basic oxides.

Non-metal oxides are acidic oxides.

KEY POINT

An important general rule is:

- **metal oxides** are **basic**
  When soluble, metal oxides form alkaline solutions in water
- **non-metal oxides** are **acidic**
  When soluble, non-metal oxides form acidic solutions in water.

The action of water on Period 3 oxides is summarised below.

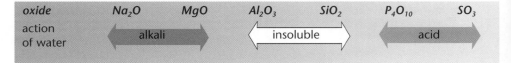

oxide	Na_2O	MgO	Al_2O_3	SiO_2	P_4O_{10}	SO_3
action of water	alkali		insoluble		acid	

Progress check

1 (a) Across the Periodic Table, what pattern is shown in the formulae of the Period 3 oxides?

(b) Why do elements in the **same** group form compounds with **similar** formulae?

2 The reactions of elements with oxygen are redox reactions. Use oxidation numbers to show that the reaction of aluminium with oxygen is a redox reaction.

1 (a) For each element in the period, there is an increase of 0.5 mole of O per mole of the element.

(b) Although the outer shell is different, there is the same number of electrons in the outer shell.

2 $4Al(s) + 3O_2(g) \longrightarrow 2Al_2O_3(s)$

Al changes oxidation number from 0 to +3 (oxidation)

O changes oxidation state from 0 to −2 (reduction)

4.2 Chlorides of the Period 3 elements

After studying this section you should be able to:

LEARNING SUMMARY

- *describe and explain the periodic variation of the formulae and boiling points of the main chlorides of Period 3*
- *describe the reactions of the Period 3 elements with chlorine*
- *describe the action of water on Period 3 chlorides*
- *describe these reactions in terms of structure and bonding*

The formulae of the Period 3 chlorides

AQA	M5	SALTERS	M5
EDEXCEL	M4	WJEC	CH5
OCR	M5.1	NICCEA	M4
NUFFIELD	M6		

Key points from AS

- **The modern Periodic Table**
  Revise AS pages 64–67
- **Redox reactions**
  Revise AS pages 68–70

The relationship of the formulae of some Period 3 chlorides with electron structure and oxidation state is shown below.

element	Na	Mg	Al	Si	P
electrons in outer shell	1	2	3	4	5
formula of chloride	$NaCl$	$MgCl_2$	Al_2Cl_6	$SiCl_4$	PCl_3 PCl_5
oxidation state of element in chloride	+1	+2	+3	+4	+3 +5

increase in highest oxidation state →

The oxidation state of the element in its chloride is equal to the number of chlorine atoms bonded to the element.

The amount of chlorine atoms per mole of the element increases steadily across Period 3.

For each element in the Period, there is an increase of 1 mole of Cl per mole of element.

$NaCl$ → $MgCl_2$ → Al_2Cl_6 → $SiCl_6$ → $SiCl_4$ → PCl_5

The same trend is seen across each Period of the Periodic Table.

The reason is the same – the number of electrons in the outer shell increases steadily across each period.

As with the Period 3 oxides (see page 64), this periodicity in formulae is also seen as a repeating trend across Periods 2 and 4. The table below shows the formulae of the chlorides from Period 2 to Period 4 of the Periodic Table.

Period 2	$LiCl$	$BeCl_2$	BCl_3	CCl_4
Period 3	$NaCl$	$MgCl_2$	$AlCl_3$	$SiCl_4$
Period 4	KCl	$CaCl_2$	$GaCl_3$	$GeCl_4$

Preparation of chlorides from the elements

AQA	M5	SALTERS	M5
EDEXCEL	M4	WJEC	CH5
OCR	M5.1	NICCEA	M4
NUFFIELD	M6		

Chlorides of Period 3 can be prepared by heating the elements in chlorine. As with the Period 3 oxides (see page 65), these are redox reactions.

Metal chlorides

Equations for the reactions of chlorine with the metals sodium, magnesium and aluminium are shown below.

$$2Na(s) + Cl_2(g) \longrightarrow 2NaCl(s)$$

$$Mg(s) + Cl_2(g) \longrightarrow MgCl_2(s)$$

$$2Al(s) + 3Cl_2(g) \longrightarrow 2Al_2Cl_6(s)$$

$NaCl$ and $MgCl_2$ are white **ionic** compounds.

Al_2Cl_6 is a white covalent compound.

At room temperature and pressure:

- $NaCl$ and $MgCl_2$ are white **ionic compounds** with **giant** structures
- Al_2Cl_6 is a white compound but, surprisingly, it has **covalent** bonding and a **simple molecular** structure.

Keys points from AS

- **Polarisation in ionic compounds**
  Revise AS page 54

PCl_5 and $SiCl_4$ are **covalent compounds.**

At room temperature and pressure:
$SiCl_4$ is a colourless liquid
PCl_5 is a white solid.

Although aluminium has the characteristic properties of a metal, its compounds often show covalent or intermediate bonding (as in Al_2O_3 on page 65). This property results from the large polarising effect of the very small Al^{3+} ion.

Non-metal chlorides

Equations for the reactions of silicon and phosphorus with chlorine are shown below.

$$Si(s) + 2Cl_2(g) \longrightarrow SiCl_4(l)$$

$$2P(s) + 5Cl_2(g) \longrightarrow 2PCl_5(s)$$

- $SiCl_4$ and PCl_5 are **covalent compounds.**

Trend in the boiling points of the Period 3 chlorides

AQA	M5	SALTERS	M5
EDEXCEL	M4	WJEC	CH5
OCR	M5.1	NICCEA	M4
NUFFIELD	M6		

The table below compares the boiling points of some Period 3 chlorides with structure and bonding.

chloride	*NaCl*	*$MgCl_2$*	*Al_2Cl_6*	*$SiCl_4$*	*PCl_5*
boiling point /°C	1413	1412	178	58	56
structure	giant lattice		simple molecules		
bonding	ionic		covalent		
forces	**strong** forces between **ions**		**weak** forces between **molecules**		

Key points from AS

- **Bonding, structure and properties**
  Revise AS pages 58–61
- **Trends in boiling points of Period 3 elements**
  Revise AS pages 66–67

Giant structures and boiling point

The chlorides with **giant** structures (NaCl $\longrightarrow$ $MgCl_2$)

- have **high boiling points**
- require a large amount of energy to break the **strong forces** between their particles
- NaCl and $MgCl_2$ have ionic bonding and the **strong** forces are electrostatic attractions acting between positive and negative **ions.**

The chlorides with **simple molecular** structures (Al_2Cl_6 $\longrightarrow$ PCl_5)

- have low boiling points
- require a small amount of energy to break the weak **van der Waals' forces** acting between the **molecules.**

Comparison between the oxides and chlorides of Period 3

The structure and bonding of the Period 3 oxides and chlorides are compared below.

For **oxides**, bonding and structure change at **different** points.

- Bonding changes from ionic to covalent between MgO and SiO_2 with Al_2O_3 having intermediate bonding.
- Structure changes from giant to simple molecular between SiO_2 and P_4O_{10}.

For **chlorides**, both bonding and structure change at the **same** point.

Between $MgCl_2$ and Al_2Cl_6:

- bonding changes from ionic to covalent **and**
- structure changes from giant to simple molecular.

The action of water on the chlorides of Period 3

AQA	M5	SALTERS	M5
EDEXCEL	M4	WJEC	CH5
OCR	M5.1	NICCEA	M4
NUFFIELD	M6		

Dissolving ionic chlorides

The ionic chlorides **dissolve** in water forming a **neutral** or very weakly acidic solution.

$$NaCl(s) + aq \longrightarrow Na^+(aq) + Cl^-(aq) \qquad pH = 7$$

$$MgCl_2(s) + aq \longrightarrow Mg^{2+}(aq) + 2Cl^-(aq) \qquad pH = 6$$

Reactions of covalent chlorides

The covalent chlorides are **hydrolysed** by water in **a vigorous reaction**. Strong acid solutions are formed containing hydrochloric acid.

Equations for the reactions of $Al_2Cl_6(s)$, $SiCl_4(l)$ and $PCl_5(s)$ with water are shown below.

$$Al_2Cl_6(s) + 6H_2O(l) \longrightarrow 2Al(OH)_3(s) + 6HCl(aq)$$

$$SiCl_4(l) + 2H_2O(l) \longrightarrow SiO_2(s) + 4HCl(aq)$$

$$PCl_5(s) + 4H_2O(l) \longrightarrow H_3PO_4(aq) + 5HCl(aq)$$

KEY POINT

An important general rule is:
- **ionic chlorides** form **neutral** solutions in water
- **covalent chlorides** form **acidic** solutions in water.

This periodicity of properties is repeated across each period.

The action of water on the Period 3 oxides is summarised below.

chloride	*NaCl*	*$MgCl_2$*	*$AlCl_3$*	*$SiCl_4$*	*PCl_5*
action of water	neutral	very weak acid	strong acid		

Progress check

1 (a) What is the trend in pH shown by the chlorides of Period 3?
(b) Water is added to sodium chloride and phosphorus trichloride to form aqueous solutions with different pH values. Explain, with equations, why these solutions have different pH values.

1 (a) neutral to acidic
(b) $NaCl(s) + aq \longrightarrow Na^+(aq) + Cl^-(aq)$
$PCl_3(s) + 3H_2O(l) \longrightarrow H_3PO_3(aq) + 3HCl(aq)$
NaCl(aq) simply dissolves in water and does not generate H^+ or OH^- ions in solution. ∴ NaCl(aq) is neutral.
PCl_3 reacts with water forming H_3PO_3 and HCl both generating H^+ ions in solution. ∴ the solution is acidic.

4.3 The transition elements

After studying this section you should be able to:

- *explain what is meant by the terms d-block element and transition element*
- *deduce the electronic configurations of atoms and ions of the d-block elements Sc → Zn*
- *know the typical properties of a transition element*

LEARNING SUMMARY

Key points from AS

- **The Periodic Table**
  Revise AS page 64–86

During AS Chemistry, you studied the metals in the s-block of the Periodic Table. For A2 Chemistry, this understanding is extended with a detailed study of the elements scandium to zinc in the d-block of the Periodic Table. Many of the principles introduced during AS Chemistry will be revisited, especially redox chemistry, and structure and bonding.

The d-block elements

AQA	M5	SALTERS	M4
EDEXCEL	M5	WJEC	CH5
OCR	M5.1	NICCEA	M5
NUFFIELD	M6		

The d-block is found in the centre of the Periodic Table. For A2 Chemistry, you will be studying mainly the d-block elements in Period 4 of the Periodic Table: Sc, Ti, V, Cr, Mn, Fe, Co, Ni, Cu and Zn.

Electronic configurations of d-block elements

The diagram below shows the sub-shell being filled across Period 4 of the Periodic Table.

4s		3d										4p					
K	Ca	Sc	Ti	V	Cr	Mn	Fe	Co	Ni	Cu	Zn	Ga	Ge	As	Se	Br	Kr

Key points from AS

- **Sub-shells and orbitals**
  Revise AS pages 21–22
- **Filling the sub-shells**
  Revise AS pages 23–24
- **Sub-shells and the Periodic Table**
  Revise AS page 24

The 3d sub-shell is at a higher energy and fills **after** the 4s sub-shell.

See also the stability from a half-full p sub-shell.

- The elements Sc ⟶ Zn have their highest energy electrons in the 4s and 3d sub-shells.
- Across the d-block, electrons are filling the 3d sub-shell. The diagram below shows the outermost electron shell for Sc ⟶ Zn.

Notice how the orbitals are filled:

- the orbitals in the 3d sub-shell are first occupied singly to prevent any repulsion caused by pairing.

Notice chromium and copper:

- chromium has one electron in each orbital of the 4s and 3d sub-shells ⟶ extra stability
- copper has a full 3d sub-shell ⟶ extra stability.

		4s	3d				
Sc	[Ar] $3d^1 4s^2$	↑↓	↑				
Ti	[Ar] $3d^2 4s^2$	↑↓	↑	↑			
V	[Ar] $3d^3 4s^2$	↑↓	↑	↑	↑		
Cr	[Ar] $3d^5 4s^1$	↑	↑	↑	↑	↑	↑
Mn	[Ar] $3d^5 4s^2$	↑↓	↑	↑	↑	↑	↑
Fe	[Ar] $3d^6 4s^2$	↑↓	↑↓	↑	↑	↑	↑
Co	[Ar] $3d^7 4s^2$	↑↓	↑↓	↑↓	↑	↑	↑
Ni	[Ar] $3d^8 4s^2$	↑↓	↑↓	↑↓	↑↓	↑	↑
Cu	[Ar] $3d^{10} 4s^1$	↑	↑↓	↑↓	↑↓	↑↓	↑↓
Zn	[Ar] $3d^{10} 4s^2$	↑↓	↑↓	↑↓	↑↓	↑↓	↑↓

Transition elements

The elements in the d-block, Ti→Cu, are called transition elements.

KEY POINT

- A transition element has **at least one ion** with a **partially filled d sub-shell.**

Although scandium and zinc are d-block elements, they are **not** *transition elements*. Neither element forms an ion with a partially filled d sub-shell.

Scandium forms one ion only, Sc^{3+} with the 3d sub-shell **empty**.

$$\text{Sc [Ar]}3d^14s^2 \xrightarrow{\textit{loss of 3 electrons}} Sc^{3+} \text{ only: } \quad \text{[Ar]}$$

Zinc forms one ion only, Zn^{2+} with the 3d sub-shell **full**.

$$\text{Zn [Ar]}3d^{10}4s^2 \xrightarrow{\textit{loss of 2 electrons}} Zn^{2+} \text{ only: } \quad \text{[Ar]}3d^{10}$$

d–block elements

Sc	Ti	V	Cr	Mn	Fe	Co	Ni	Cu	Zn

transition elements (Ti to Cu)

Key points from AS

- **The s-block elements**
  Revise AS pages 71–75

Properties of transition elements

1 high density
2 high m. and b. pt.
3 moderate to low reactivity
4 two or more oxidation states
5 coloured ions
6 complex ions with ligands
7 catalysts

Typical properties of transition elements

The transition elements are all metals – they are good conductors of heat and electricity. In addition, some general properties distinguish the transition elements from the s-block metals.

Density

- The transition elements are more **dense** than other metals.
  - They have smaller atoms than the metals in Groups 1 and 2. The small atoms are able to pack closely together ⟶ high density.

Melting point and boiling point

- The transition elements have **higher melting and boiling points** than other metals.
  - Within the metallic lattice, ions are **smaller** than those of the s-block metals – the metallic bonding between the metal ions and the delocalised electrons is strong.

Reactivity

- The transition elements have **moderate to low reactivity**.
  - Unlike the s-block metals, they do **not** react with cold water. Many transition metals react with dilute acids although some, such as gold, silver and platinum, are extremely unreactive.

Oxidation states

- Most transition elements have compounds with **two or more oxidation states**.

Colour of compounds

- The transition elements have **coloured compounds and ions**.
  - Most transition metals have at least one oxidation state that is coloured.

Complex ions

- The ions of transition elements form **complex ions** with ligands.

Catalysis

- Many transition elements can act as **catalysts**.

Formation of ions

AQA	M5	SALTERS	M4
EDEXCEL	M5	WJEC	CH5
OCR	M5.1	NICCEA	M5
NUFFIELD	M6		

The energy levels of the 4s and 3d sub-shells are very close together.

KEY POINT

- Transition elements are able to form **more than one ion**, each with a different oxidation state, by losing the **4s** electrons and different numbers of **3d** electrons.
- When forming ions, the 4s electrons are **lost first**, before the 3d electrons.

The Fe^{3+} ion, [Ar]$3d^5$ is more stable than the Fe^{2+} ion, [Ar]$3d^5 4s^1$.

Fe^{3+} has a half-full 3d sub-shell ⟶ stability.

Forming Fe^{2+} and Fe^{3+} ions from iron

Iron has the electronic configuration [Ar]$3d^6 4s^2$

Iron forms two common ions, Fe^{2+} and Fe^{3+}.

An Fe atom loses **two 4s electrons** ⟶ Fe^{2+} ion.

An Fe atom loses **two 4s electrons** and **one 3d electron** ⟶ Fe^{3+} ion.

The 4s and 3d energy levels are very close together and both are involved in bonding.

When forming ions, the 4s electrons are **lost before** the 3d electrons.

Note that this is different from the order of **filling** – 4s before 3d.

Once the 3d sub-shell starts to fill, the 4s electrons are repelled to a higher energy level and are lost first.

The variety of oxidation states

The table below summarises the stable oxidation states of the elements scandium to zinc. Those in bold type represent the commonest oxidation numbers of the elements in their compounds.

Sc	*Ti*	*V*	*Cr*	*Mn*	*Fe*	*Co*	*Ni*	*Cu*	*Zn*
	+1	+1	+1	+1	+1	+1	+1	**+1**	
	+2	+2	+2	**+2**	**+2**	**+2**	**+2**	**+2**	**+2**
+3	+3	+3	**+3**	+3	**+3**	+3	+3	+3	
	+4	+4	+4	**+4**	+4	+4	+4		
		+5	+5	+5	+5	+5			
			+6	+6	+6				
				+7					

- The variety of oxidation states results from the close similarity in energy of the 4s and the 3d electrons.
- The number of available oxidation states for the element **increases** from Sc → Mn.
  - All of the available 3d and 4s electrons may be used for bond formation.
- The number of available oxidation states for the element **decreases** from Mn → Zn.
  - There is a decreasing number of unpaired d electrons available for bond formation.
- Higher oxidation states involve covalency because of the high charge densities involved.
  - In practice, for oxidation states of +4 and above, the bonding electrons are involved in covalent bond formation.

Refer to the electronic configurations of the d-block elements, shown on p. 70.

Progress check

1 Write down the electronic configuration of the following ions:
(a) Ni^{2+} (b) Cr^{3+} (c) Cu^{+} (d) V^{3+}.

2 What is the oxidation state of the metal in each of the following:
(a) MnO_2 (b) CrO_3 (c) MnO_4^- (d) VO_2^+?

1 (a) [Ar]$3d^8$ (b) [Ar]$3d^3$ (c) [Ar]$3d^{10}$ (d) [Ar].
2 (a) +4 (b) +6 (c) +7 (d) +5.

4.4 Transition element complexes

After studying this section you should be able to:

LEARNING SUMMARY

- *explain what is meant by the terms 'complex ion' and 'ligand'*
- *predict the formula and possible shape of a complex ion*
- *know that ligands can be unidentate, bidentate and multidentate*
- *know that transition metal ions can be identified by their colour*
- *know that electronic transitions are responsible for colour*
- *explain how colorimetry can be used to determine the concentration and formula of a complex ion*

Ligands and complex ions

AQA	M5	SALTERS	M4
EDEXCEL	M5	WJEC	CH5
OCR	M5.1	NICCEA	M5
NUFFIELD	M6		

Transition metal ions are small and densely charged. They strongly attract electron-rich species called **ligands** forming **complex ions**.

A ligand has a lone pair of electrons.

KEY POINT

A ligand is a molecule or ion that bonds to a metal ion by:
- forming a coordinate (dative covalent) bond
- donating a lone pair of electrons into a vacant d-orbital.

Common ligands include: H_2O**:**, **:**Cl^-, **:**NH_3, **:**CN^-

KEY POINT

A **complex ion** is a central metal ion surrounded by ligands.
The **coordination number** is the total **number** of **coordinate bonds** from ligands to the central transition metal ion of a complex ion.

Key points from AS

- **Electron pair repulsion theory**
  Revise AS pages 50–51

For AS, you learnt that electron pair repulsion is responsible for the shape of a molecule or ion.

Shapes of complex ions

An important factor in deciding the geometry of the complex ion is the size of the ligand.

Six coordinate complexes

Six water molecules are able to fit around a Cu^{2+} ion to form:
- the complex ion $[Cu(H_2O)_6]^{2+}$
- with a coordination number of **6**.

The **six** electron pairs surrounding the central Cu^{2+} ion in $[Cu(H_2O)_6]^{2+}$ repel one another as far apart as possible forming a complex ion with a an **octahedral** shape.

The hexaaquacopper(II) complex ion, $[Cu(H_2O)_6]^{2+}$

$[Cu(H_2O)_6]^{2+}$
octahedral
6 coordinate

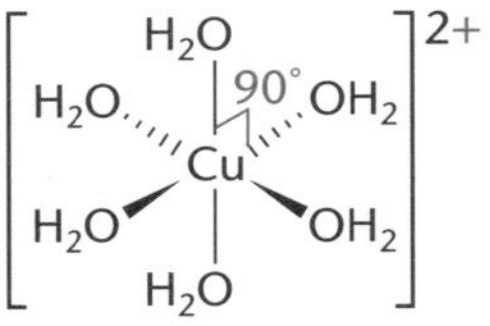

The size of a ligand helps to decide the geometry of the complex ion.

Four coordinate complexes

Chloride ions are larger than water molecules and it is only possible for **four** chloride ions to fit around the central copper(II) ion to form:
- the complex ion $[CuCl_4]^{2-}$
- with a coordination number of **4**.

The **four** electron pairs surrounding the central Cu^{2+} ion in $[CuCl_4]^{2-}$ repel one another as far apart as possible to form a complex ion with a **tetrahedral** shape.

The tetrachlorocuprate(II) complex ion $[CuCl_4]^{2-}$

Notice the overall charge on the complex ion $[CuCl_4]^{2-}$ is 2–.
Cu^{2+} and 4 Cl^- → 2– ions.

$[CuCl_4]^{2-}$
tetrahedral
4 coordinate

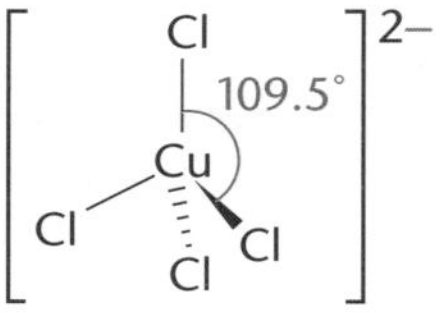

General rules for deciding the shape of a complex ion

Although there are exceptions, the following general rules are useful.

> **KEY POINT**
>
> Complex ions with small ligands such as H_2O and NH_3 are usually **6-coordinate** and **octahedral**.
>
> Complex ions with large Cl^- ligands are usually **4-coordinate** and **tetrahedral**.

The table below compares complex ions formed from cobalt(II) and iron(III) ions.

ligand	*complex ions from Co^{2+}*	*complex ions from Fe^{3+}*	*shape*
H_2O	$[Co(H_2O)_6]^{2+}$	$[Fe(H_2O)_6]^{3+}$	octahedral
NH_3	$[Co(NH_3)_6]^{2+}$	$[Fe(NH_3)_6]^{3+}$	octahedral
Cl^-	$[CoCl_4]^{2-}$	$[FeCl_4]^-$	tetrahedral

- Some complex ions contain more than one type of ligand. For example, copper(II) forms a complex ion with a mixture of water and ammonia ligands, $[Cu(NH_3)_4(H_2O)_2]^{2+}$.
- Ag^+ forms linear complexes that are 2-coordinate: e.g. $[Ag(H_2O)_2]^+$, $[Ag(NH_3)_2]^+$ and $[AgCl_2]^-$.

Ligands have teeth

AQA M5 | SALTERS M4 | OCR M5.6 | NICCEA M5 | NUFFIELD M6

Ligands such as H_2O, NH_3 and Cl^- are called **unidentate** ligands. They form only **one** coordinate bond to the central metal ion.

A molecule or ion with more than one oxygen or nitrogen atom may form more than one coordinate bond to the central metal ion.

Monodentate means **one** tooth.
Bidentate ligands have **two** teeth.
Multidentate ligands, **many** teeth.
Each 'tooth' is a coordinate bond.

Ligands that can form **two** coordinate bonds to the central metal ion are called **bidentate** ligands. Examples of bidentate ligands are ethane-1,2-diamine, $NH_2CH_2CH_2NH_2$ and the ethanedioate ion, $(COO^-)_2$.

Multidentate ligands can form **many** coordinate bonds to the central metal ion. For example, the **hexa**dentate ligand, $edta^{4-}$, is able to form **six** coordinate bonds to the central metal ion. The diagram of $edta^{4-}$ shows four oxygen atoms and two nitrogen atoms able to form coordinate bonds.

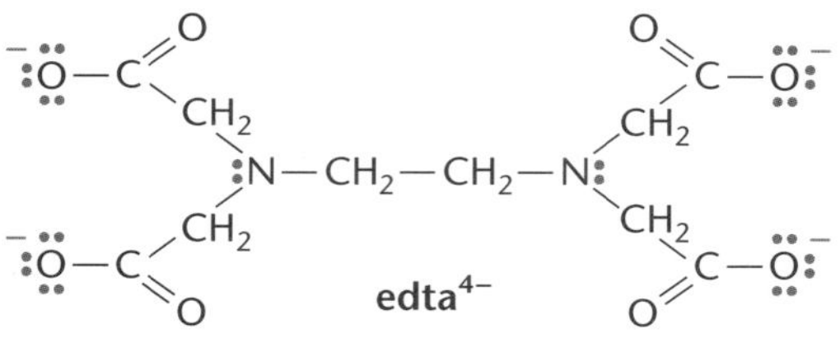

Ligands as Lewis bases

AQA M5

A Lewis acid is an electron pair acceptor
A Lewis base is an electron pair donor

A ligand behaves as:
- a **Lewis base** by **donating** an electron pair to the central metal ion of a complex ion.

The transition metal ion acts as:
- a **Lewis acid** by **accepting** the electron pair from the ligand.

These definitions of an acid and a base extend acid–base chemistry to reactions which do not involve transfer of protons.

Colour and complex ions

AQA M5 | EDEXCEL M5 | OCR M5.1 | NUFFIELD M6 | SALTERS M4 | WJEC CH5 | NICCEA M5

Complex ions have different colours

Many transition metal ions have different colours. These colours indicate the possible identity of a transition metal ion. The table below shows the colours of some **aqua** complex ions of d-block elements. Notice that different oxidation states of a d-block element can have different colours.

Sc	Ti	V	Cr	Mn	Fe	Co	Ni	Cu	Zn
								+1 *colourless*	
				+2 *pale pink*	+2 *pale green*	+2 *pink*	+2 *green*	+2 *blue*	+2 *colourless*
+3 *colourless*	+3 *violet*	+3 *blue*	+3 *ruby*		+3 *red-brown*				
		+5 yellow							
			+6 *yellow or orange*						
				+7 *purple*					

Why are transition element ions coloured?

AQA M5 | EDEXCEL M5 | OCR M5.6 | WJEC CH5

Splitting the 3d energy level

All five d orbitals in the outer d sub-shell of an **isolated** transition element ion have the same energy. However, when ligands bond to the central ion, the d orbitals move to two different energy levels. The gap between the energy levels depends upon:

- the ligand
- the coordination number
- the transition metal ion.

This diagram shows how water ligands split the energy levels of the five 3d orbitals in a copper(II) ion.

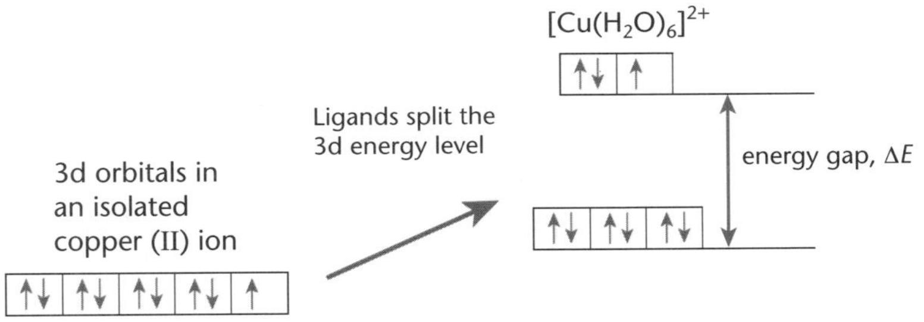

Absorption of light energy

An electron can be promoted from the lower 3d energy level by absorbing energy **exactly** equal to the energy gap ΔE. This energy is provided by radiation from the visible and ultraviolet regions of the spectrum.

Energy and frequency

The difference in energy between the d-orbitals, ΔE, and the frequency of the absorbed radiation, f, are linked by the following relationship:

$\Delta E = hf$

h = Planck's constant, 6.63 x 10^{-34} J s.

Radiation with a frequency f has an energy hf, where h = Planck's constant, 6.63 × 10^{-34} J s.

So to promote an electron through as energy gap ΔE, a quantity of light energy is needed given by:

$$\Delta E = hf$$

The diagram below shows what happens when a complex ion absorbs light energy.

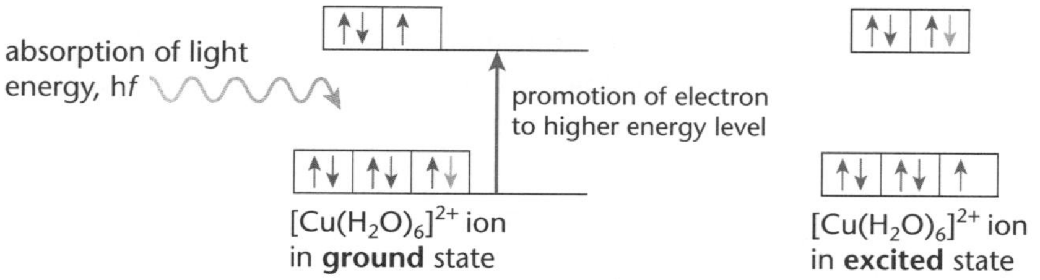

Only those ions that have a partially filled d subshell are coloured. Cu^{2+} is [Ar]$3d^9 4s^2$

The blue colour of $[Cu(H_2O)_6]^{2+}$ results from:

- **absorption** of light energy in the red, yellow and green regions of the spectrum. This absorbed radiation provides the energy for an electron to be excited to a higher energy level
- **reflection** of blue light only, giving the copper(II) hexaaqua ion its characteristic blue colour. The blue region of the spectrum is **not** absorbed.

Copper has another ion, Cu^+, with the electronic configuration: [Ar]$3d^{10}$.

The Cu^+ ion is colourless because:

- the 3d sub-shell is full, [Ar]$3d^{10}$, preventing electron transfer.

Ag^+ complexes are also colourless because the 3d sub-shell is full.

KEY POINT

The colour of a transition metal complex ion results from the **transfer** of an **electron** between the orbitals of an **unfilled** d sub-shell.

Using colorimetry

AQA	M5	SALTERS	M4
OCR	M5.1	WJEC	CH5
NUFFIELD	M4, M6		

A colorimeter measures the amount of light absorbed by a coloured solution. The concentration and formula of a complex ion can be determined from the intensity of light absorbed by the colorimeter.

Finding an unknown concentration

A calibration curve for the colorimeter is first plotted using solutions of a complex ion of known concentration.

The solution of the complex ion with an unknown concentration is then placed in the colorimeter and the absorbance measured. The concentration can be simply determined from the calibration curve.

Progress check

1 Predict the formula of the following complex ions:
(a) iron(III) with water ligands
(b) vanadium(III) with chloride ligands
(c) nickel(II) with ammonia ligands.

2 Explain the origin of colour in a complex ion that is yellow.

1 (a) $[Fe(H_2O)_6]^{3+}$ (b) $[VCl_4]^-$ (c) $[Ni(NH_3)_6]^{2+}$.

2 The 3d orbitals are split between two energy levels. A 3d electron from the lower 3d energy level absorbs visible light in the blue and red regions of the spectrum. The absorbed energy allows the d electron to be promoted to a vacancy in the orbitals of the higher 3d energy level. With the red and blue regions of the spectrum absorbed, only yellow light is reflected, giving the complex ion its colour.

4.5 Ligand substitution of complex ions

After studying this section you should be able to:

- *describe what is meant by ligand substitution*
- *understand that ligand exchange may produce changes in colour and coordination number*
- *understand stability of complex ions in terms of stability constants*

LEARNING SUMMARY

Exchange between ligands

AQA	M5	SALTERS	M4
EDEXCEL	M5	WJEC	CH5
OCR	M5.1	NICCEA	M5
NUFFIELD	M6		

Ligand substitution usually produces a change in colour.

A **ligand substitution** reaction takes place when a ligand in a complex ion exchanges for another ligand. A change in ligand usually changes the energy gap between the 3d energy levels. With a different ΔE value, light with a different frequency is absorbed, producing a colour change.

Exchange between H_2O and NH_3 ligands

Remember that the **size** of the ligand is a major factor in deciding the coordination number (see p. 73):

- H_2O and NH_3 have similar sizes – same coordination number
- H_2O and Cl^- have different sizes – different coordination numbers.

The **similar sizes** of water and ammonia molecules ensure that ligand exchange takes place with **no change in coordination number**.

For example, addition of an excess of concentrated aqueous ammonia to aqueous nickel(II) ions results in ligand substitution of ammonia ligands for water ligands. Both complex ions have an octahedral shape with a coordination number of six.

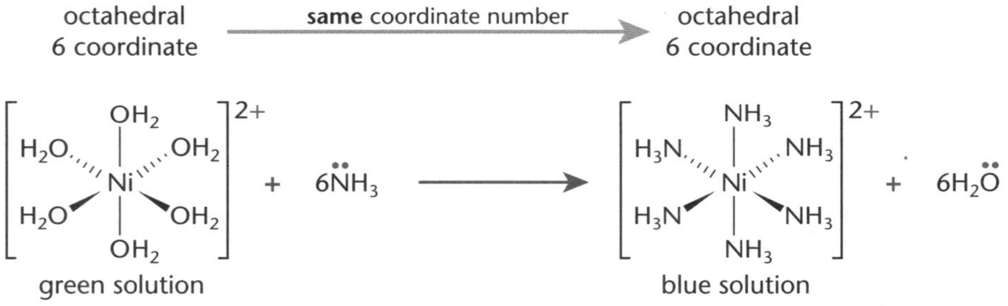

$$[Ni(H_2O)_6]^{2+} + 6NH_3 \longrightarrow [Ni(NH_3)_6]^{2+} + 6H_2O$$

Exchange between H_2O and Cl^- ligands

You should also be able to construct a balanced equation for any ligand exchange reaction.

The **different sizes** of water molecules and chloride ions ensure that ligand exchange takes place **with a change in coordination number**.

For example, addition of an excess of concentrated hydrochloric acid (as a source of Cl^- ions) to aqueous cobalt(II) ions results in ligand substitution of chloride ligands for water ligands. The complex ions have different shapes with different coordination numbers.

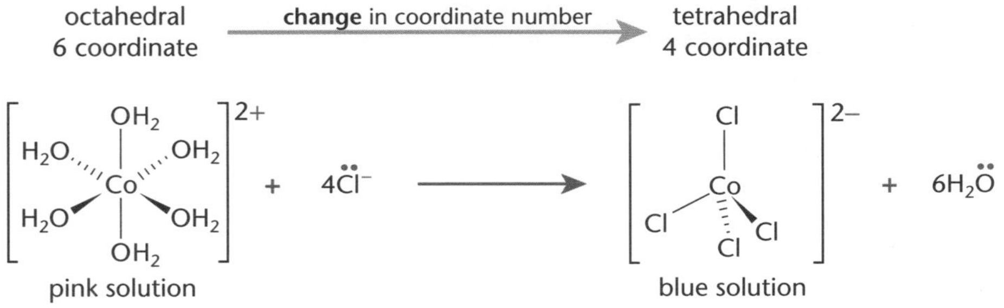

$$[Co(H_2O)_6]^{2+} + 4Cl^- \longrightarrow [CoCl_4]^{2-} + 6H_2O$$

Ligand substitutions of copper(II) and cobalt(II)

Concentrated hydrochloric acid is used as a source of Cl^- ligands.

Concentrated aqueous ammonia is used as a source of NH_3 ligands.

Ligand substitutions of copper(II) and cobalt(II) complexes are shown below.

Copper

$[Cu(H_2O)_6]^{2+}$ blue octahedral

- HCl(aq) → $[CuCl_4]^{2-}$ yellow tetrahedral
- NH_3(aq) → $[Cu(NH_3)_4(H_2O)_2]^{2+}$ deep blue octahedral

Cobalt

$[Co(H_2O)_6]^{2+}$ pink octahedral

- HCl(aq) → $[CoCl_4]^{2-}$ blue tetrahedral
- NH_3(aq) → $[Co(NH_3)_6]^{2+}$ pale straw octahedral

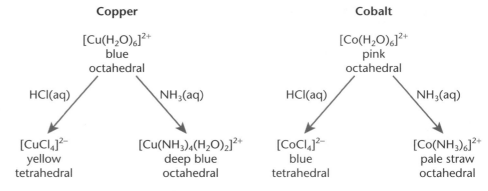

Incomplete substitution

This reaction is used as a test for the Fe^{3+} ion – the deep red-brown colour of $[Fe(H_2O)_5(SCN)]^{2+}$ is intense and extremely small traces of Fe^{3+} ions can be detected using this ligand substitution reaction.

Substitution may be incomplete.

- Aqueous ammonia only exchanges **four** from the six water ligands in $[Cu(H_2O)_6]^{2+}$.

$$[Cu(H_2O)_6]^{2+} + 4NH_3 \longrightarrow [Cu(NH_3)_4(H_2O)_2]^{2+} + 4H_2O$$

blue solution → deep blue solution

- Aqueous thiocyanate ions, SCN^-, only exchanges **one** from the six water ligands in $[Fe(H_2O)_6]^{3+}$.

$$[Fe(H_2O)_6]^{3+} + SCN^-(aq) \longrightarrow [Fe(H_2O)_5(SCN)]^{2+} + H_2O$$

pale brown solution → deep red solution

Stability of complex ions

NUFFIELD M6 | NICCEA M5 | SALTERS M4

The concentration of water is virtually constant. Constant terms are not included in expressions for equilibrium constants.

The stability of a complex ion depends upon the ligands. For example, a complex ion of copper(II) is more stable with ligands of ammonia than with water ligands.

The stability of a complex ion is measured as a **stability constant**, K_{stab}:

$$[Cu(H_2O)_6]^{2+} + 4NH_3(aq) \rightleftharpoons [Cu(NH_3)_4(H_2O)_2]^{2+} + 4H_2O(l)$$

$$K_{stab} = \frac{[Cu(NH_3)_4(H_2O)_2]^{2+}}{[Cu(H_2O)_6]^{2+}\,[NH_3(aq)]^4} = 1 \times 10^{12}\ dm^{12}\ mol^{-4} \text{ at } 25°C$$

Stability constants usually measure the stability of a complex compared with the metal aqua-ion. The greater the value of K_{stab}, the more stable the complex ion.

The data is expressed on a logarithmic scale because of the very large range of values.

complex ion	***log K_{stab}***
$[Fe(H_2O)_5(SCN)]^{2+}$	3.9
$[CuCl_4]^{2-}$	5.6
$[Ni(NH_3)_6]^{2+}$	8.6
$[Ni(edta)]^{2-}$	19.3

↓ increased stability

Progress check

1 Write equations for the following ligand substitutions:
(a) $[Ni(H_2O)_6]^{2+}$ with six ammonia ligands
(b) $[Fe(H_2O)_6]^{3+}$ with four chloride ligands
(c) $[Cr(H_2O)_6]^{3+}$ with six cyanide ligands, CN^-.

1 (a) $[Ni(H_2O)_6]^{2+} + 6NH_3 \longrightarrow [Ni(NH_3)_6]^{2+} + 6H_2O$
(b) $[Fe(H_2O)_6]^{3+} + 4Cl^- \longrightarrow [FeCl_4]^- + 6H_2O$
(c) $[Cr(H_2O)_6]^{3+} + 6CN^- \longrightarrow [Cr(CN)_6]^{3-} + 6H_2O$.

4.6 Redox reactions of transition metal ions

After studying this section you should be able to:

- *describe redox behaviour in transition elements*
- *construct redox equations, using relevant half-equations*
- *perform calculations involving simple redox titrations*

LEARNING SUMMARY

Redox reactions

AQA	M5	SALTERS	M4
EDEXCEL	M5	WJEC	CH5
OCR	M5.1	NICCEA	M5
NUFFIELD	M6		

Transition elements have variable oxidation states with characteristic colours (see page 75). Many **redox** reactions take place in which transition metals ions change their oxidation state by **gaining** or **losing** electrons.

Key points from AS

- **Redox reactions**
  Revise AS pages 68–70

The iron(II) – manganate(VII) reaction

Iron(II) ions are oxidised by acidified manganate(VII) ions.

- Acidified manganate(VII) ions are reduced to manganate(II) ions.

reduction: $MnO_4^-(aq) + 8H^+(aq) + 5e^- \longrightarrow Mn^{2+}(aq) + 4H_2O(l)$
+7 → +2
purple ⟶ very pale pink

- Iron(II) ions are oxidised to iron(III) ions.

oxidation: $Fe^{2+}(aq) \longrightarrow Fe^{3+}(aq) + e^-$
+2 → +3

The reaction is easy to see.
The deep purple MnO_4^- ions are reduced to the very pale pink Mn^{2+} ions, which are virtually colourless in solution.

To give the overall equation:

- the electrons are balanced

$$MnO_4^-(aq) + 8H^+(aq) + 5e^- \longrightarrow Mn^{2+}(aq) + 4H_2O(l)$$
$$5Fe^{2+}(aq) \longrightarrow 5Fe^{3+}(aq) + 5e^-$$

- the half-equations are added

$$5Fe^{2+}(aq) + MnO_4^-(aq) + 8H^+(aq) \longrightarrow 5Fe^{3+}(aq) + Mn^{2+}(aq) + 4H_2O(l)$$

Redox titrations

AQA	M5	SALTERS	M4
EDEXCEL	M5	WJEC	CH5
OCR	M5.1	NICCEA	M5
NUFFIELD	M6		

Redox titrations can be used in analysis.

Essential requirements are:

- an oxidising agent
- a reducing agent
- an easily-seen colour change (with or without an indicator).

Redox titrations using the acidified manganate(VII)

- $KMnO_4$ is reacted with a reducing agent such as Fe^{2+} under acidic conditions (see above).
- Purple $KMnO_4$ is added from the burette to a measured amount of the acidified reducing agent.
- At the end-point, the colour changes from colourless to pale pink, showing that all the reducing agent has exactly reacted. The pale-pink colour indicates the first trace of an excess of purple manganate(VII) ions.
- This titration is **self-indicating** – no indicator is required, the colour change from the reduction of MnO_4^- ions being sufficient to indicate that reaction is complete.

It is important to use an acid that does **not** react with either the oxidising or reducing agent – dilute sulphuric acid is usually used.
For example, hydrochloric acid cannot be used because HC is oxidised to Cl_2 by MnO_4^-.

The commonest redox titration encountered at A Level is that between manganate(VII) and acidified iron(II) ions.

Key points from AS

- **Calculations in acid–base titrations**
  Revise AS page 40

Calculations for redox titrations follow the same principles as those used for acid–base titrations, studied during AS Chemistry.

Example

Five iron tablets with a combined mass of 0.900 g were dissolved in acid and made up to 100 cm^3 of solution. In a titration, 10.0 cm^3 of this solution reacted exactly with 10.4 cm^3 of 0.0100 mol dm^{-3} potassium manganate(VII). What is the percentage by mass of iron in the tablets?

As with all titrations, we must consider **five** pieces of information:

- the balanced equation
  $5Fe^{2+}(aq) + MnO_4^-(aq) + 8H^+(aq) \rightarrow 5Fe^{3+}(aq) + Mn^{2+}(aq) + 4H_2O(l)$
- the concentration c_1 and reacting volume V_1 of $KMnO_4(aq)$
- the concentration c_2 and reacting volume V_2 of $Fe^{2+}(aq)$.

The concentration and volume of $KMnO_4$ are known so the amount of MnO_4^- ions can be found.

From the titration results, the amount of $KMnO_4$ can be calculated:

$$\text{amount of } KMnO_4 = c \times \frac{V}{1000} = 0.0100 \times \frac{10.4}{1000} = 1.04 \times 10^{-4} \text{ mol}$$

Using the equation determine the number of moles of the second reagent.

From the equation, the amount of Fe^{2+} can be determined:

$5Fe^{2+}(aq) + MnO_4^-(aq) + 8H^+(aq) \rightarrow 5Fe^{3+}(aq) + Mn^{2+}(aq) + 4H_2O(l)$

5 mol **1 mol** *(balancing numbers)*

$\therefore$ 5 x 1.04 × 10^{-4} mol Fe^{2+} reacts with 1.04 × 10^{-4} mol MnO_4^-

$\therefore$ amount of Fe^{2+} that reacted = 5.20 × 10^{-4} mol

If there is sampling of the original solution, you will need to scale.

Find the amount of Fe^{2+} in the solution prepared from the tablets:

10.0 cm^3 of $Fe^{2+}(aq)$ contains 5.20 × 10^{-4} mol Fe^{2+} (aq)
The 100 cm^3 solution of iron tablets contains 10 × (5.20 × 10^{-4})
= 5.20 × 10^{-3} mol Fe^{2+}

Find the percentage of Fe^{2+} in the tablets (A_r: Fe, 55.8)

5.20 x 10^{-3} mol Fe^{2+} has a mass of 5.20 × 10^{-3} × 55.8 = 0.290 g

$$\therefore \text{ \% of } Fe^{2+} \text{ in tablets} = \frac{\text{mass of } Fe^{2+}}{\text{mass of tablets}} \times 100 = \frac{0.290}{0.900} \times 100 = 32.2\%$$

Further redox titrations

Providing there is a visible colour change, many other redox reactions can be used for redox titrations.

For example, acidified dichromate(VI) ions is an oxidising agent which can also oxidise Fe^{2+} ions to Fe^{3+}.

reduction: $Cr_2O_7^{2-}(aq) + 14H^+(aq) + 6e^- \longrightarrow 2Cr^{3+}(aq) + 7H_2O(l)$
oxidation: $Fe^{2+}(aq) \longrightarrow Fe^{3+}(aq) + e^-$
overall: $6Fe^{2+}(aq) + Cr_2O_7^{2-}(aq) + 14H^+(aq) \rightarrow 6Fe^{3+}(aq) + 2Cr^{3+}(aq) + 7H_2O(l)$

Further examples of redox reactions

AQA M5 | OCR M5.6 | EDEXCEL M5 | WJEC M5

Vanadium

Vanadium(V) can be reduced to vanadium(II) using zinc in acid solution as the reducing agent. The formation of different oxidation states of vanadium can be followed by colour changes.

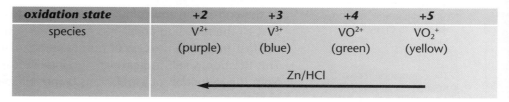

oxidation state	**+2**	**+3**	**+4**	**+5**
species	V^{2+} (purple)	V^{3+} (blue)	VO^{2+} (green)	VO_2^+ (yellow)
	← Zn/HCl (from +5 to +2)			

Chromium

The oxidation states of chromium can be interchanged using:

- hydrogen peroxide in alkaline solution as the oxidising agent
- zinc in acid solution as the reducing agent.

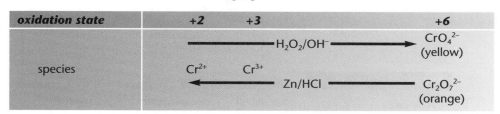

oxidation state	**+2**	**+3**		**+6**
species			H_2O_2/OH^- ⟶	CrO_4^{2-} (yellow)
	Cr^{2+}	Cr^{3+}		
			⟵ Zn/HCl	$Cr_2O_7^{2-}$ (orange)

This conversion can be explained by applying le Chatelier's principle.

Addition of acid increases [H+(aq] – the equilibrium moves to the right, counteracting this change.

Added alkali reacts with H+(aq) ions present in the equilibrium system. [H+(aq] decreases – the equilibrium moves to the left, counteracting this change.

The dichromate(VI) – chromate(VI) conversion

The table above shows that the dichromate(VI) ion $Cr_2O_7^{2-}$ and the chromate(VI) ion CrO_4^{2-} have the **same** oxidation state. These ions can be interconverted at different pH values.

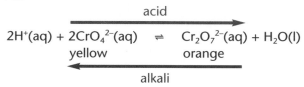

$$\underset{\text{yellow}}{2H^+(aq) + 2CrO_4^{2-}(aq)} \underset{\text{alkali}}{\overset{\text{acid}}{\rightleftharpoons}} \underset{\text{orange}}{Cr_2O_7^{2-}(aq) + H_2O(l)}$$

Cobalt

Cobalt(II) can be oxidised to Co(III) using:

- hydrogen peroxide in alkaline solution or
- air in the presence of ammonia as an oxidising agent.

oxidation state	**+2**		**+3**
species	$[Co(NH_3)_6]^{2+}$ (light straw)	H_2O_2/OH^- *or* NH_3/air ⟶	$[Co(NH_3)_6]^{3+}$ (brown)

Useful oxidising and reducing agents

KEY POINT

For moving **up** oxidation states:

- H_2O_2/OH^- is a good general oxidising agent.

For moving **down** oxidation states:

- Zn/HCl is a good general reducing agent.

Progress check

1 Write a full equation from the following pairs of half-equations. For each pair, identify the changes in oxidation number, what has been oxidised and what has been reduced.

(a) $Zn \longrightarrow Zn^{2+} + 2e^-$

$VO_2^+ + 4H^+ + 3e^- \longrightarrow V^{2+} + 2H_2O$

(b) $Cr^{3+} + 8OH^- \longrightarrow CrO_4^{2-} + 4H_2O + 3e^-$

$H_2O_2 + 2e^- \longrightarrow 2OH^-$

2 In a redox titration, 25.0 cm^3 of an acidified solution containing Fe^{2+}(aq) ions reacted exactly with 21.8 cm^3 of 0.0200 mol dm^{-3} potassium manganate(VII). Calculate the concentration of Fe^{2+}(aq) ions.

1 (a) $2VO_2^+ + 3Zn + 8H^+ \longrightarrow 2V^{2+} + 3Zn^{2+} + 4H_2O$

V: +5 → +2; Zn: 0 → +2; VO_2^+ reduced; Zn oxidised.

(b) $2Cr^{3+} + 3H_2O_2 + 10OH^- \longrightarrow 2CrO_4^{2-} + 8H_2O$

Cr: +3 → +6; O: −1 → −2; Cr^{3+} oxidised; H_2O_2 reduced.

2 0.0872 mol dm^{-3}

4.7 Catalysis

After studying this section you should be able to:

- *understand how a transition element acts as a catalyst*
- *explain how a catalyst acts in heterogeneous catalysis*
- *explain how a catalyst acts in homogeneous catalysis*

LEARNING SUMMARY

Key points from AS

- **How do catalysts work?** *Revise AS page 102*
- **Heterogeneous and homogeneous catalysis** *Revise AS pages 102–103*

The two different classes of catalyst, homogeneous and heterogeneous, were discussed in detail during AS Chemistry. Both types are used industrially but heterogeneous catalysis is much more common. The AS coverage is summarised below and expanded to include extra examples of transition element catalysis.

Transition elements as catalysts

AQA	M5	SALTERS	M4
EDEXCEL	M5	WJEC	CH5
OCR	M5.1	NICCEA	M5
NUFFIELD	M6		

An important use of transition metals and their compounds is as catalysts for many industrial processes. Nickel and platinum are extensively used in the petroleum and polymer industries.

> Transition metal ions are able to act as catalysts by changing their oxidation states. This is made possible by using the partially full d orbitals for gaining or losing electrons.
>
> KEY POINT

Heterogeneous catalysis

AQA	M5	SALTERS	M4
EDEXCEL	M5	WJEC	CH5
OCR	M5.6	NICCEA	M5

A **heterogeneous** catalyst has a **different** phase from the reactants.

- Many examples of heterogeneous catalysts involve reactions between **gases**, catalysed by a **solid** catalyst which is often a transition metal or one of its compounds.

Key points from AS

- **Heterogeneous and homogeneous catalysis** *Revise AS pages 102–103*

The process involves

- **diffusion** of gas molecules onto the surface of the iron catalyst
- **adsorption** of the gases to the surface of the catalyst
- **weakening of bonds**, allowing a chemical reaction to take place
- **diffusion** of the product molecules from the surface of the catalyst, allowing more gas molecules to diffuse onto its surface.

Choice of a heterogeneous catalyst (AQA only)

A key requirement of a heterogeneous catalyst is to bind reactant molecules to the surface of the catalyst by adsorption.

The strength of this adsorption is critical.

- Tungsten, W, bonds very strongly to reactant molecules. The adsorption is so strong that molecules are unable to leave the surface and the catalyst is poisoned and rendered useless.
- Silver, Ag, bonds too weakly to reactant molecules to allow a reaction to take place.

Catalysts can become coated by impurities and can become 'poisoned' – efficiency is reduced as active sites are blocked.

Nickel and platinum are commonly used as heterogeneous catalysts:

- the bonding of reactant molecules is sufficient to allow a reaction to take place but
- weak enough to allow molecules to escape following reaction.

Transition metal ions as heterogeneous catalysts

The catalytic converter

- Rh/Pt/Pd catalysts are used in catalytic converters for the removal of polluting gases, such as CO and NO, produced in a car engine.
- The **solid** Rh/Pt/Pd catalyst is supported on a ceramic honeycomb, giving a large surface area for the catalyst which can be spread extremely thinly on the support. The high surface area of catalyst increases the chances of a reaction and less of the expensive catalysts are needed, keeping down costs.

Iron

Iron-containing catalysts are used in many industrial processes, the most important being to form ammonia from nitrogen and hydrogen in the Haber process.

Vanadium

A vanadium(V) oxide, V_2O_5, catalyst is used in the Contact process for sulphur trioxide production from which sulphuric acid is manufactured.

This catalysis proceeds via an **intermediate state**.

Notice that the oxidation number of each atom has been shown. Two V atoms have been reduced from +5 to +4.

- The vanadium(V) oxide catalyst first oxidises sulphur dioxide to sulphur trioxide. The vanadium is **reduced** from the +5 to +4 oxidation state, forming vanadium(IV) oxide $V_2O_4(s)$ as the **intermediate state**.

$$SO_2(g) + V_2O_5(s) \longrightarrow SO_3(g) + V_2O_4(s) \qquad \textit{vanadium reduced}$$

$$+4 \longrightarrow +6$$

$$+5 \longrightarrow +4$$

$$+5 \longrightarrow +4$$

- The intermediate V_2O_4 is then oxidised back to the +5 oxidation state by oxygen, forming vanadium(V) oxide.

$$\tfrac{1}{2}O_2(g) + V_2O_4(s) \longrightarrow V_2O_5(s) \qquad \textit{vanadium oxidised}$$

$$0 \longrightarrow -2$$

$$+4 \longrightarrow +5$$

$$+4 \longrightarrow +5$$

This is a chain reaction because the vanadium(V) oxide is then able to oxidise further sulphur dioxide.

Note

- Although V_2O_5 has been involved in the mechanism, the overall reaction does not include V_2O_5.
  - V_2O_5 is unchanged at the end of the reaction.
- The equation for the overall reaction taking place is:

$$SO_2(g) + \tfrac{1}{2}O_2(g) \rightleftharpoons SO_3(g) \qquad \textit{The Contact process}$$

- The vanadium is able to catalyse the reaction by interchanging its +5 and +4 oxidation states.

Homogeneous catalysis

AQA	M5	WJEC	CH5
EDEXCEL	M5	NICCEA	M5
NUFFIELD	M6		

Key points from AS

- **How do catalysts work**
  Revise AS page 102

A **homogeneous** catalyst has the **same** phase as the reactants.

- During homogeneous catalysis, the catalyst initially reacts with the reactants forming an **intermediate state** with **lower activation energy**.
- The catalyst is then regenerated as the reaction completes, allowing the catalyst to react with further reactants.
- Overall the catalyst is **not** used up, but it is actively involved in the process by a chain reaction.

Transition metal ions as homogeneous catalysts

The reaction between I^- and $S_2O_8^{2-}$, catalysed by $Fe^{2+}(aq)$ ions

The reaction between I^- and $S_2O_8^{2-}$ is very slow:

- it can be catalysed by Fe^{2+} ions
- the Fe^{2+} ions are recycled in a chain reaction.

chain reaction: $S_2O_8^{2-}(aq) + 2Fe^{2+}(aq) \longrightarrow 2SO_4^{2-}(aq) + 2Fe^{3+}(aq)$

$2Fe^{3+}(aq) + 2I^-(aq) \longrightarrow I_2(aq) + 2Fe^{2+}(aq)$

overall: $S_2O_8^{2-}(aq) + 2I^-(aq) \longrightarrow I_2(aq) + 2SO_4^{2-}(aq)$

The overall reaction between I^- and $S_2O_8^{2-}$ involves two negative ions. These negative ions will repel one another, making it difficult for a reaction to take place. Notice that, in each stage of the mechanism, the catalyst provides positive ions which attract, and react with, each negative ion in turn.

Notice that, as in the first example, this reaction is between two negative ions.
The mechanism of this reaction proceeds in a similar manner – the catalyst provides positive ions which react with each negative ion.

Autocatalysis by $Mn^{2+}(aq)$ ions

In warm acidic conditions, manganate(VII) ions, MnO_4^-, oxidise ethanedioate ions, $C_2O_4^{2-}$ to carbon dioxide.

$$2MnO_4^-(aq) + 5C_2O_4^{2-}(aq) + 8H^+(aq) \rightarrow 2Mn^{2+}(aq) + 10CO_2(g) + H_2O(l)$$

The purple manganate(VII) colour disappears very slowly at first, indicating a slow reaction. As soon as $Mn^{2+}(aq)$ ions start to form, the colour disappears immediately because the $Mn^{2+}(aq)$ ions catalyse the reaction. This is called **autocatalysis**.

Progress check

1 (a) State what is meant by homogeneous and heterogeneous catalysis.
(b) State an example of a homogeneous and heterogeneous catalyst.

2 How does a transition metal act as a catalyst?

1 (a) Homogeneous: catalyst and reactants have the same phase.
Heterogeneous: catalyst and reactants have different phases.
(b) Homogeneous: Fe^{2+} catalysing the reaction between I^- and $S_2O_8^{2-}$.
Heterogeneous: iron catalysing the reaction between N_2 and H_2 for ammonia production (Haber process).

2 Transition metal changes oxidation states by gaining or losing electrons from partially filled d orbitals.

4.8 Reactions of metal aqua-ions

LEARNING SUMMARY

After studying this section you should be able to:

- *describe the precipitation reactions of metal aqua-ions with bases*
- *describe the acidity of transition metal aqua-ions*
- *describe the acid–base behaviour of metal hydroxides*

Simple precipitation reactions of metal aqua-ions

AQA	M5	SALTERS	M4
EDEXCEL	M5	WJEC	CH5
OCR	M5.1	NICCEA	M5
NUFFIELD	M6		

Notice that NaOH(aq) and NH_3(aq) **both** act as bases ⟶ precipitate of the transition metal hydroxide.

A precipitation reaction takes place between **aqueous alkali** and an aqueous solution of a **metal(II)** or **metal(III) cation**.

This results in the formation of a **precipitate** of the **metal hydroxide**, often with a characteristic colour.

Suitable aqueous alkalis include aqueous sodium hydroxide, NaOH(aq), and aqueous ammonia, NH_3(aq).

Aqueous ammonia provides OH^-(aq) ions.
$NH_3(aq) + H_2O(l) \rightleftharpoons NH_4^+(aq) + OH^-(aq)$

These precipitation reactions can be represented simply as follows.

$$Cu^{2+}(aq) + 2OH^-(aq) \longrightarrow Cu(OH)_2(s)$$
$$Cr^{3+}(aq) + 3OH^-(aq) \longrightarrow Cr(OH)_3(s)$$

The hydroxides of all transition metals are insoluble in water.

The characteristic colour of the precipitate can help to identify the metal ion. The colours of some hydroxide precipitates are shown below.

hydroxide	***$Fe(OH)_2(s)$***	***$Fe(OH)_3(s)$***	***$Co(OH)_2(s)$***	***$Cu(OH)_2(s)$***	***$Cr(OH)_3(s)$***
colour	green	brown	turquoise	blue	green

Reaction of complex metal aqua-ions with aqueous alkali

Equations can be written using complex aqua-ions for the reactions above, each producing a **precipitate** of the **hydrated hydroxide**.

The precipitate has no charge.

$$[Cu(H_2O)_6]^{2+}(aq) + 2OH^-(aq) \longrightarrow Cu(OH)_2(H_2O)_4(s) + 2H_2O(l)$$
blue precipitate

$$[Cr(H_2O)_6]^{3+}(aq) + 3OH^-(aq) \longrightarrow Cr(OH)_3(H_2O)_3(s) + 3H_2O(l)$$
green precipitate

Excess aqueous sodium hydroxide

Metal(III) hydroxides **dissolve** in an **excess** of dilute aqueous sodium hydroxide forming charged **anions**.

Species in solution are charged.

$$Al(OH)_3(H_2O)_3(s) + OH^-(aq) \rightleftharpoons [Al(OH)_4]^-(aq) + 3H_2O(l)$$
aluminate ions

$$Cr(OH)_3(H_2O)_3(s) + 3OH^-(aq) \rightleftharpoons [Cr(OH)_6]^{3-}(aq) + 3H_2O(l)$$
chromate(III) ions

Excess aqueous ammonia

Excess ammonia usually results in **ligand exchange** forming a soluble ammine complex:

$$Cr(H_2O)_3(OH)_3(s) + 6NH_3(aq) \longrightarrow [Cr(NH_3)_6]^{3+}(aq) + 3H_2O(l) + 3OH^-(aq)$$
green solution

If aqueous ammonia is slowly added to an aqueous solution of a **metal(II)** or **metal(III) cation**:

- a precipitate of the metal hydroxide first forms: **acid–base reaction**
- the precipitate then dissolves in excess ammonia: **ligand substitution**.

Progress check

In exams, a great emphasis is placed on these reactions. You will need to learn the colours of all solutions and precipitates in your course.

1 Write equations for the following reactions of metal aqua-ions with aqueous alkali. What is the colour of each precipitate?
(a) $Co^{2+}(aq)$ (b) $Fe^{2+}(aq)$ (c) $[Ni(H_2O)_6]^{2+}$ (d) $[Fe(H_2O)_6]^{3+}$.

2 Write equations to show what happens when excess $NH_3(aq)$ is added to the following hydrated metal hydroxides.
(a) $Co(OH)_2(H_2O)_4$ (b) $Cu(OH)_2(H_2O)_4$ (c) $Cr(OH)_3(H_2O)_3$.

1 (a) $Co^{2+}(aq) + 2OH^-(aq) \longrightarrow Co(OH)_2(s)$ (blue-green)
(b) $Fe^{2+}(aq) + 2OH^-(aq) \longrightarrow Fe(OH)_2(s)$ (green)
(c) $[Ni(H_2O)_6]^{2+}(aq) + 2OH^-(aq) \longrightarrow Ni(OH)_2(H_2O)_4(s) + 2H_2O(l)$ (blue-green)
(d) $[Fe(H_2O)_6]^{3+}(aq) + 3OH^-(aq) \longrightarrow Fe(OH)_3(H_2O)_3(s) + 3H_2O(l)$ (brown)

2 (a) $Co(OH)_2(H_2O)_4(s) + 6NH_3(aq) \longrightarrow [Co(NH_3)_6]^{2+}(aq) + 4H_2O(l) + 2OH^-(aq)$ (pale straw)
(b) $Cu(OH)_2(H_2O)_4(s) + 4NH_3(aq) \longrightarrow [Cu(NH_3)_4(H_2O)_2]^{2+}(aq) + 2H_2O(l) + 2OH^-(aq)$ (deep-blue)
(c) $Cr(OH)_3(H_2O)_3(s) + 6NH_3(aq) \longrightarrow [Cr(NH_3)_6]^{3+}(aq) + 3H_2O(l) + 3OH^-(aq)$ (green)

Precipitation as acid–base equilibria

AQA	M5	WJEC	CH5
EDEXCEL	M5	NICCEA	M5

Acidity of metal aqua-ions

Fission of O–H bond

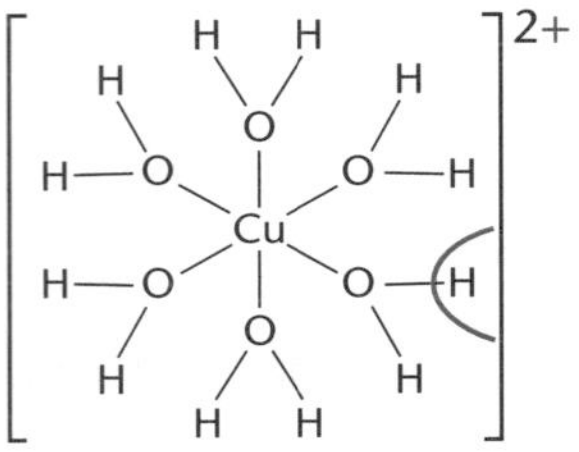

Metal(II) and metal(III) aqua-ions behave as weak acids. This is achieved by:

- fission of the O–H bond in one of the water ligands of the complex ion
- donation of a proton to a water molecule.

Metal(II) cations

Metal(II) aqua-ions are weaker acids than metal(III) aqua-ions.

- With a metal(II) aqua-ion, the equilibrium lies well to the left-hand side: only a very weak acid is formed.

$$[Cu(H_2O)_6]^{2+}(aq) + H_2O(l) \rightleftharpoons [Cu(H_2O)_5(OH)]^{+}(aq) + H_3O^{+}(l)$$

very weakly acidic

Metal(III) cations

Examples of metal aqua-ions with 3+ cations include $[Al(H_2O)_6]^{3+}$, $[V(H_2O)_6]^{3+}$, $[Cr(H_2O)_6]^{3+}$ and $[Fe(H_2O)_6]^{3+}$.

- Metal(III) aqua-ions are slightly more acidic than metal(II) aqua-ions: the equilibrium position has moved slightly to the right.

$$[Fe(H_2O)_6]^{3+}(aq) + H_2O(l) \rightleftharpoons [Fe(H_2O)_5(OH)]^{2+}(aq) + H_3O^{+}(aq)$$

weakly acidic

KEY POINT: The aqua-ions of metal(III) cations have a greater charge/size ratio and greater acidity than aqua-ions of metal(II) cations.

Precipitation as acid–base equilibria

The precipitation reaction of a transition metal with hydroxide ions can be expressed as a series of acid–base equilibria.

The equilibrium set up in water by the hexaaqua ion of iron(III) is:

$$[Fe(H_2O)_6]^{3+}(aq) + H_2O(l) \rightleftharpoons [Fe(H_2O)_5(OH)]^{2+}(aq) + H_3O^{+}(aq)$$

On addition of aqueous hydroxide ions:

OH^- reacts with H_3O^+:
$OH^- + H_3O^+ \longrightarrow 2H_2O$

- H_3O^+ ions are removed from the equilibrium, which shifts to the right.

The $[Fe(H_2O)_5(OH)]^{2+}$ ion now sets up a second equilibrium:

$$[Fe(H_2O)_5(OH)]^{2+}(aq) + H_2O(l) \rightleftharpoons [Fe(H_2O)_4(OH)_2]^{+}(aq) + H_3O^{+}(aq)$$

- Again, the hydroxide ion removes H_3O^+ ions from the aqueous equilibrium, which shifts to the right.
- This process continues. Eventually **all charge** will be **removed** from the iron(III) complex. Then the uncharged **hydrated metal hydroxide** precipitates.

$$[Fe(H_2O)_4(OH)_2]^{+}(aq) + H_2O(l) \rightleftharpoons [Fe(H_2O)_3(OH)_3](s) + H_3O^{+}(aq)$$

With the weak alkali $Na_2CO_3(aq)$, metal(II) aqua-ions precipitate metal **carbonates.**

The more acidic metal(III) aqua-ions precipitate metal **hydroxides.**

Metal(III) carbonates do not exist.

Acid–base properties of metal hydroxides

AQA	M5	WJEC	CH5
EDEXCEL	M5	NICCEA	M5

Many insoluble metal hydroxides are **amphoteric** – they can act as both an acid and a base.

Metal hydroxides as acids

See also: 'Reaction of complex metal aqua-ions with aqueous alkali' p. 85.

Some insoluble metal hydroxides can act as acids, donating protons to strong alkalis and forming soluble anions.

- Metal(III) hydroxides react with dilute aqueous sodium hydroxide.

$$Al(OH)_3(H_2O)_3(s) + OH^-(aq) \rightleftharpoons [Al(OH)_4]^-(aq) + 3H_2O(l)$$

aluminate ions

$$Cr(OH)_3(H_2O)_3(s) + 3OH^-(aq) \rightleftharpoons [Cr(OH)_6]^{3-}(aq) + 3H_2O(l)$$

chromate(III) ions

In general, metal(II) hydroxides form anions with the general formula $[M(OH)_4]^{-2}$.
Metal(III) hydroxides form anions with the general formula $[M(OH)_6]^{3-}$.

- Metal(II) hydroxides are weaker acids and react only with concentrated aqueous sodium hydroxide. E.g.

$$Cu(OH)_2(H_2O)_4(s) + 2OH^-(aq) \rightleftharpoons [Cu(OH)_4]^{2-}(aq) + 4H_2O(l)$$

deep blue cuprate(II) ions

Metal hydroxides as bases (AQA only)

All metal hydroxides can act as bases, removing protons from strong acids.

$$Cu(OH)_2(H_2O)_4(s) + 2H^+(aq) \rightleftharpoons [Cu(H_2O)_6]^{2+}(aq) + 2H_2O(l)$$

blue precipitate

$$Cr(OH)_3(H_2O)_3(s) + 3H^+(aq) \rightleftharpoons [Cr(H_2O)_6]^{3+}(aq) + 3H_2O(l)$$

green precipitate

This reaction reverses the direction of the acid-base equilibria that produce hydrated hydroxide precipitates (see p. 86)

Summary of reactions of metal aqua-ions with bases

AQA	M5	SALTERS	M4
EDEXCEL	M5	WJEC	CH5
OCR	M5.1	NICCEA	M5
NUFFIELD	M6		

	base				
	OH⁻(aq)		***NH₃(aq)***		***CO₃²⁻(aq)***
ion		***excess***		***excess***	
$[Fe(H_2O)_6]^{2+}$ green	$Fe(OH)_2(s)$ green	no change	$Fe(OH)_2(s)$ green	no change	$FeCO_3(s)$ white
$[Fe(H_2O)_6]^{3+}$ pale violet	$Fe(OH)_3(s)$ brown	no change	$Fe(OH)_3(s)$ brown	no change	$Fe(OH)_3(s)$ brown
$[Co(H_2O)_6]^{2+}$ pink	$Co(OH)_2(s)$ blue-green	$[Co(OH)_4]^{2-}(aq)$ deep blue	$Co(OH)_2(s)$ blue-green	$[Co(NH_3)_6]^{2+}(aq)$ pale straw	$CoCO_3(s)$ pink
$[Cu(H_2O)_6]^{2+}$ blue	$Cu(OH)_2(s)$ pale blue	$[Cu(OH)_4]^{2-}(aq)$ deep blue	$Cu(OH)_2(s)$ pale blue	$[Cu(NH_3)_4(H_2O)_2]^{2+}(aq)$ deep blue	$CuCO_3(s)$ green
$[Al(H_2O)_6]^{3+}$ colourless	$Al(OH)_3(s)$ white	$[Al(OH)_4]^-(aq)$ colourless	$Al(OH)_3(s)$ white	no change	$Al(OH)_3(s)$ white
$[Cr(H_2O)_6]^{3+}$ ruby	$Cr(OH)_3(s)$ green	$[Cr(OH)_6]^{3-}(aq)$ green	$Cr(OH)_3(s)$ green	$[Cr(NH_3)_6]^{3+}(aq)$ purple	$Cr(OH)_3(s)$ green

- The precipitate of $Fe(OH)_2(s)$ is oxidised in air forming a brown precipitate of $Fe(OH)_3(s)$.
- The complexion $[Co(NH_3)_6]^{2+}(aq)$ darkens in air as it is oxidised to $[Co(NH_3)_6]^{3+}(aq)$ (see page 81).

Sample question and model answer

(a) Scandium and zinc both appear in the d-block of the Periodic Table but their common compounds do not show the characteristic properties associated with transition element compounds.

Give the electronic configurations of each of these elements and deduce the oxidation state each has in its common compounds. [4]

Scandium [Ar] $3d^14s^2$ ✓ Zinc [Ar] $3d^{10}4s^2$ ✓
Scandium has +3 oxidation state. ✓ Zinc has +2 oxidation state. ✓

(b) When concentrated hydrochloric acid is added to an aqueous solution of a cobalt(II) salt there is a change in colour. Account for this observation by giving the two cobalt species involved, their colours, their shapes and the co-ordination numbers of the cobalt they contain. Write an equation for the reaction and explain why the addition of dilute hydrochloric acid does not cause the same colour change.

This is standard bookwork but it also reinforces important principles:

H_2O forms octahedral complexes with a coordination number of 6

Cl^- forms tetrahedral complexes with a coordination number of 4.

Aqueous cobalt(II) contains $[Co(H_2O)_6]^{2+}$ ✓ which is pink. ✓ It has an octahedral shape ✓ and a coordination number of 6. ✓
After addition of concentrated hydrochloric acid, the complex ion $[CoCl_4]^{2-}$ ✓ is formed which is blue. ✓ It has an tetrahedral shape ✓ and a coordination number of 4. ✓

Equation: $[Co(H_2O)_6]^{2+} + 4Cl^- \longrightarrow CoCl_4^{2-} + 6H_2O$ ✓✓

The concentration of chloride ligands in dilute hydrochloric acid is much smaller than the concentration of water ligands. ✓ [11]

(c) A 0.570 g sample of hydrated iron(II) sulphate was dissolved in water. After acidification with dilute sulphuric acid, the solution was found to react exactly with 22.8 cm^3 of 0.0180 M potassium manganate(VII) solution.

(i) Write half-equations for the oxidation of Fe^{2+} and the reduction of MnO_4^- and deduce the equation for the overall reaction.

Learn these important equations.

If your equation is wrong, you can still score in the next part provided that your method is built upon sound chemical principles.

Show your working!!

$MnO_4^-(aq) + 8H^+(aq) + 5e^- \longrightarrow Mn^{2+}(aq) + 4H_2O(l)$ ✓
$Fe^{2+}(aq) \longrightarrow Fe^{3+}(aq) + e^-$ ✓
$5Fe^{2+}(aq) + MnO_4^-(aq) + 8H^+(aq) \longrightarrow Mn^{2+}(aq) + 4H_2O(l)$ ✓

(ii) Use the above data to calculate the relative molecular mass of hydrated iron(II) sulphate. Hence calculate the number of moles of water of crystallisation present in one mole of hydrated iron(II) sulphate.

amount of $KMnO_4 = c \times \frac{V}{1000} = 0.0180 \times \frac{22.8}{1000} = 4.10 \times 10^{-4}$ mol ✓

From the equation, the amount of Fe^{2+} = 5 x amount of $KMnO_4$.

$\therefore$ amount of Fe^{2+} that reacted $= 5 \times 4.10 \times 10^{-4}$ mol $= 2.05 \times 10^{-3}$ mol ✓

$\therefore \frac{0.570}{2.05 \times 10^{-3}}$ g of hydrated iron(II) sulphate contains 1 mol Fe^{2+} ✓

M_r of hydrated iron(II) sulphate = 278 ✓

M_r of $FeSO_4 = 55.8 + 32.1 + 16 \times 4 = 151.9$ ✓

Number of waters of crystallisation $= \frac{278 - 151.9}{18} = \frac{126.1}{18} = 7$ ✓

(iii) State how the volume of potassium manganate(VII) solution used would have differed if the solution of iron(II) sulphate had been left for some time before being titrated. Explain your answer.

Less $KMnO_4$ solution would have been needed ✓ because some of the Fe^{2+} will have been oxidised by air to Fe^{3+}. ✓ [11]

[Total: 26]

NEAB Equilibria and Inorganic Chemistry Q6(a–c) June 1998

Practice examination questions

1

(a) Complete the following table.

	Na	Mg	Al	Si
Formula of anhydrous chloride				

[2]

(b) (i) Write equations, including state symbols, for the changes that take place when these chlorides are added to water:

sodium chloride magnesium chloride

aluminium chloride silicon chloride. [4]

(ii) Interpret these changes on the basis of the bonding in the chlorides. [2]

(iii) Suggest why the chloride of carbon, CCl_4, does not react with water. [4]

(c) Aluminium hydroxide is amphoteric. Write two ionic equations to show the meaning of this statement. [2]

[Total: 14]

Edexcel Specimen Unit Test 4 Q4 2000

2

(a) Write equations to show what happens when the following oxides are added to water and predict approximate values for the pH of the resulting solutions.

(i) sodium oxide

(ii) sulphur dioxide. [4]

(b) What is the relationship between bond type in the oxides of the Period 3 elements and the pH of the solutions which result from addition of the oxides to water? [2]

(c) Write equations to show what happens when the following chlorides are added to water and predict approximate values for the pH of the resulting solutions.

(i) magnesium chloride

(ii) silicon tetrachloride. [4]

[Total: 10]

Assessment and Qualifications Alliance Specimen Unit Test 5 Q4 2000

3

(a) Explain how a bond is formed between a metal ion and a ligand in a complex ion. [2]

(b) Consider the complex compound $[Co(NH_2CH_2CH_2NH_2)_3]Cl_3$

(i) Name the ligand bonded to cobalt.

(ii) What name is given to this type of ligand?

(iii) What is the oxidation state of cobalt in this compound?

(iv) What is the co-ordination number of cobalt in this compound?

(v) Deduce the likely shape around cobalt in this complex. [5]

(c) From your knowledge of the reaction of cobalt(II) ions with ammonia, outline how you would prepare a solution of the complex $[Co(NH_2CH_2CH_2NH_2)_3]Cl_3$ starting from solid cobalt(II) chloride. [3]

[Total: 10]

Assessment and Qualifications Alliance Further Inorganic Chemistry Q1 June 1998

Practice examination questions (continued)

4

(a) Give the electronic configuration of a cobalt atom and a Co^{2+} ion. [2]

(b) Cobalt can be described both as a d-block element and as a transition element. State what is meant by each of the terms. [2]

(c) In aqueous solution water combines with the Co^{2+} ion to form the complex ion $[Co(H_2O)_6]^{2+}$ which gives a pink colour to the solution.

(i) What feature of the water molecule allows it to form a complex ion with Co^{2+}? [1]

(ii) What types of bond are present in the complex ion $[Co(H_2O)_6]^{2+}$? [2]

(iii) Suggest the shape of the ion $[Co(H_2O)_6]^{2+}$. [1]

(d) Consider the following reactions:

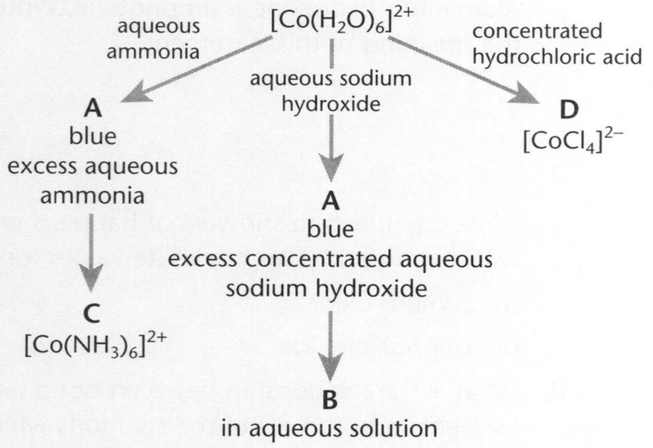

(i) Give the name of the blue precipitate, **A**, and write an ionic equation for its formation from $[Co(H_2O)_6]^{2+}$. [2]

(ii) What name is given to the type of reaction occurring in (i)? [1]

(iii) Suggest a formula for the cobalt complex ion **B** present in the solution. [1]

(iv) Give the name of the ion **C**. [1]

(v) Write an equation for the formation of ion **D** from $[Co(H_2O)_6]^{2+}$ and suggest the type of reaction taking place. [2]

[Total: 15]

Edexcel Module Test 1 Q1 modified Jan 1999

5

The concentration of hydrogen peroxide in a solution can be determined by titrating an acidified solution against aqueous potassium manganate(VII) added from a burette. The potassium manganate(VII) reacts with the colourless aqueous solution of hydrogen peroxide as shown in the equation given below.

$$5H_2O_2 + 2MnO_4^- + 6H^+ \longrightarrow 5O_2 + 8H_2O + 2Mn^{2+}$$

(a) State the role of hydrogen peroxide in this reaction. [1]

(b) Identify a suitable acid for use in this titration. [1]

(c) State the colour change at the end-point of this reaction. [1]

(d) After acidification with a suitable acid, 25.0 cm^3 of a dilute aqueous solution of hydrogen peroxide were found to react with 18.1 cm^3 of 0.0200 M $KMnO_4$. Calculate the molar concentration of hydrogen peroxide in the solution. [4]

[Total: 7]

Assessment and Qualification Alliance Equilibria and Inorganic Chemistry Q6 March 1998

Chapter 5

Isomerism, aldehydes, ketones and carboxylic acids

The following topics are covered in this chapter:

- Isomerism and functional groups
- Aldehydes and ketones
- Carboxylic acids
- Esters
- Acylation

5.1 Isomerism and functional groups

After studying this section you should be able to:

LEARNING SUMMARY

- understand what is meant by structural isomerism and stereoisomerism
- explain the term chiral centre and identify any chiral centres in a molecule of given structural formula
- understand that many natural compounds are present as one optical isomer only
- understand that many pharmaceuticals are chiral drugs
- recognise common functional groups

Key points from AS

- **Basic concepts**
  Revise AS pages 113–117

During AS Chemistry, you learnt about the basic concepts used in organic chemistry. You should revise these thoroughly before you start the A2 part of this course.

Isomerism

AQA	M4	SALTERS	M4
EDEXCEL	M4	WJEC	CH4
OCR	M4	NICCEA	M4
NUFFIELD	M6		

Structural and *cis-trans* isomerism were introduced during AS Chemistry. These are reviewed below and *cis-trans* isomerism is discussed in the wider context of **stereoisomerism**.

Key points from AS

- **Structural isomerism**
  Revise AS pages 114–115
- ***cis-trans* isomerism**
  Revise AS page 122

Structural isomerism

Structural isomers are molecules with the same molecular formula but with different structural arrangements of atoms (structural formulae).

Two structural isomers of $C_2H_4O_2$ are shown below:

Stereoisomerism

Stereoisomers are molecules with the same structural formula but their atoms have different positions in space.

There are two types of stereoisomerism, each arising from a different structural feature:

- ***cis-trans*** isomerism about a **C=C** double bond
- **optical** isomerism about a **chiral** carbon centre.

cis-trans (or geometric) isomerism

cis-trans isomerism occurs in molecules with:

- a C=C double bond and
- two **different** groups attached to **each** carbon in the C=C bond.

The double bond prevents rotation.

E.g. *cis-trans* isomers of 1,2-dichloroethene, ClCH=CHCl.

> Each *cis-trans* isomer has the same structural formula.

Optical isomers

> **KEY POINT**
>
> Optical isomerism occurs in the molecules of a compound with:
> - a **chiral** (or *asymmetric*) carbon atom.
>
> A chiral carbon atom has **four** different groups attached to it.
> - Optical isomers are mirror images of one another.

E.g. the optical isomers of an amino acid, $RCHNH_2COOH$ (see pages 123–125).

> Optical isomers are usually drawn as 3D diagrams. It is then easy to picture the mirror images.

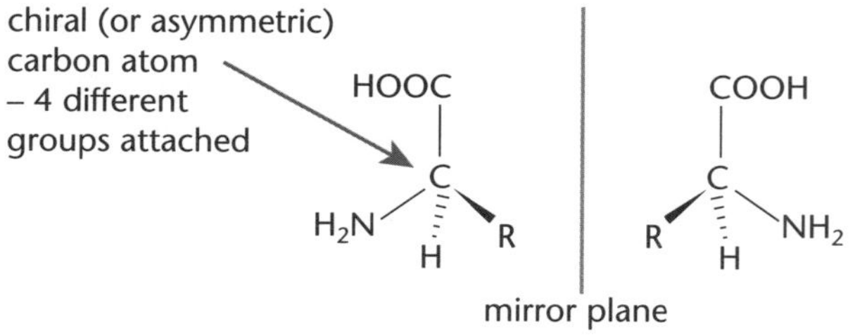

> Optical isomers are also called **enantiomers**.

Optical isomers

- are non-superimposable mirror images of one another
- rotate plane-polarised light in opposite directions
- are chemically identical.

Chirality and drug synthesis

> Society discovered the consequences of harmful side-effects from the 'wrong' optical isomer with the use of thalidomide. One optical isomer combated the effects of morning sickness in pregnant women. The other optical isomer was the cause of deformed limbs of unborn babies.
>
> Partly through this lesson, drugs are now often used as the optically pure form, comprising just the required optical isomer.

A synthetic amino acid, made in the laboratory, is optically inactive:

- it contains equal amounts of each optical isomer – a **racemic** mixture.

A natural amino acid, made by living systems, is optically active:

- it contains only one of the optical isomers.

The difference between the optical isomers present in natural and synthetic organic molecules has important consequences for drug design. The synthesis of pharmaceuticals often requires the production of chiral drugs containing a single optical isomer. Although one of the optical isomers may have beneficial effects, the other may be harmful and may lead to undesirable side effects.

Progress check

1 Show the alkenes that are structural isomers of C_4H_8.

2 Show the *cis-trans* isomers of C_4H_8.

3 Show the optical isomers of C_4H_9OH.

1 $CH_3CH_2CH{=}CH_2$; $CH_3CH{=}CHCH_3$; $(CH_3)_2C{=}CH_2$

2

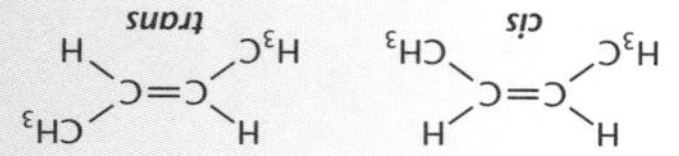

3

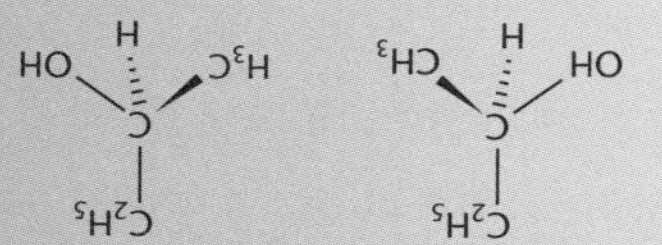

Common functional groups

AQA	M4	SALTERS	M4, M5
EDEXCEL	M4	WJEC	CH4
OCR	M4	NICCEA	M4
NUFFIELD	M4, M6		

It is essential that you can instantly identify a functional group within a molecule so that you can apply the relevant chemistry.

You must learn all of these!

The functional groups in the table below contain most of the functional groups you will meet in the A Level Chemistry course, including those met in AS Chemistry. It cannot be stressed too much just how important it is that you can instantly recognise a functional group. Without this, your progress in Organic Chemistry will be very limited.

name	*functional group*	*examples structural formula*		*prefix or suffix (for naming)*
alkane	C–H	$CH_3CH_2CH_3$ *propane*	$CH_3CH_2CH_3$	-ane
alkene	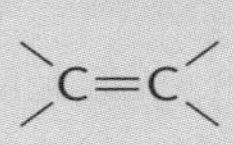	CH_3CHCH_2 *propene*	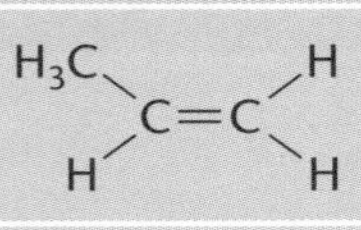	-ene
halogenoalkane	— Br	CH_3CH_2Br *bromoethane*	CH_3CH_2—Br	bromo-
alcohol	— OH	CH_3CH_2OH *ethanol*	CH_3CH_2—OH	-ol
aldehyde	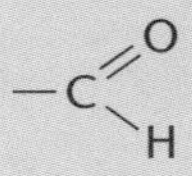	CH_3CHO *ethanal*	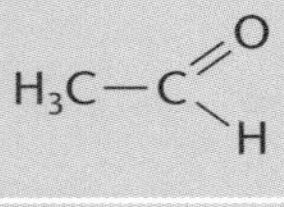	-al
ketone	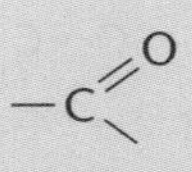	CH_3COCH_3 *propanone*	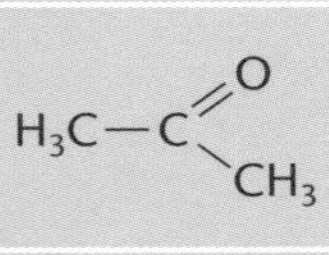	-one
carboxylic acid	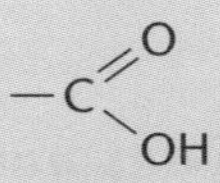	CH_3COOH *ethanoic acid*	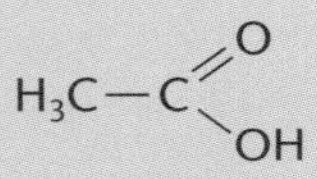	-oic acid
ester	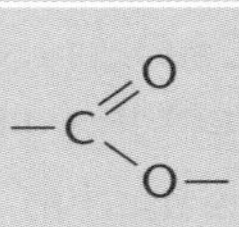	CH_3COOCH_3 *methyl ethanoate*	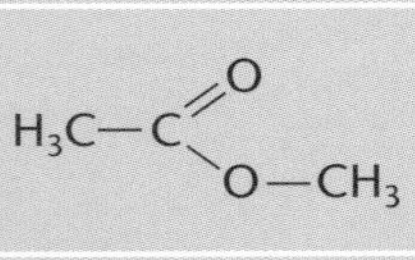	-oate
acyl chloride	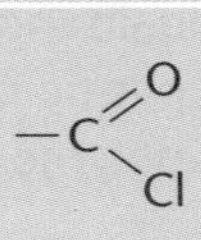	CH_3COCl *ethanoyl chloride*	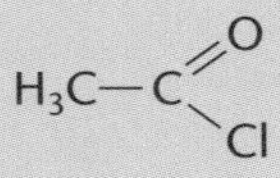	–oyl chloride
amine	— NH_2	$CH_3CH_2NH_2$ *ethylamine*	CH_3CH_2—NH_2	-amine
amide	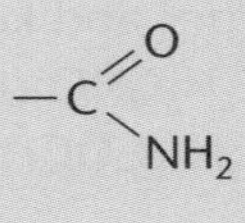	CH_3CONH_2 *ethanamide*	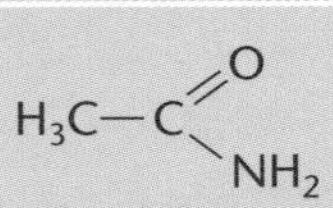	-amide
nitrile	—C≡N	CH_3CN *ethanenitrile*	H_3C—CN	-nitrile

Key points from AS

- **Functional groups**
  Revise AS pages 115–116

The nitrile carbon atom is included in the name.

CH_3CN contains a longest carbon chain with **two** carbon atoms and its name is based upon ethane – hence ethanenitrile.

5.2 Aldehydes and ketones

After studying this section you should be able to:

- *understand the polarity and physical properties of carbonyl compounds*
- *describe nucleophilic addition to aldehydes and ketones*
- *describe a test to detect the presence of the carbonyl group*
- *describe tests to detect the presence of an aldehyde group*
- *describe the reduction of carbonyl compounds to form alcohols*

LEARNING SUMMARY

Carbonyl compounds General formula: $C_nH_{2n}O$

AQA	M4	SALTERS	M5
EDEXCEL	M4	WJEC	CH4
OCR	M4	NICCEA	M4
NUFFIELD	M4		

Key points from AS

- **Oxidation of alcohols**
Revise AS pages 128–129

During AS Chemistry, you learnt about how alcohols can be oxidised to carbonyl compounds: aldehydes and ketones. For A2 Chemistry, you will learn about the reactions of aldehydes and ketones.

Types and naming of carbonyl compounds

The carbonyl group, C=O, is the functional group in aldehydes and ketones.

aldehyde, RCHO

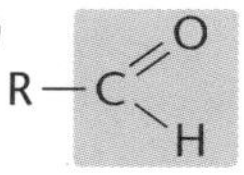

ketone, RCOR'

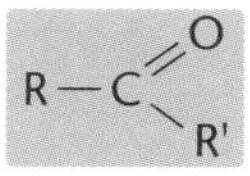

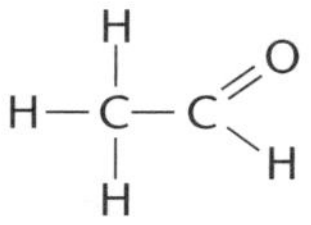

ethanal
CH_3CHO

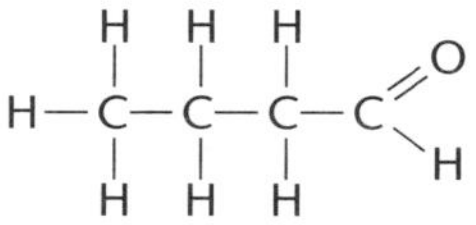

butanal
$CH_3CH_2CH_2CHO$

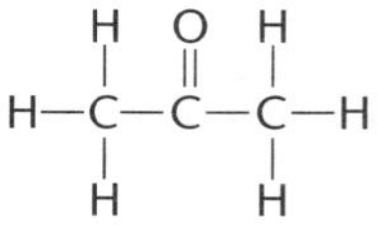

propanone
CH_3COCH_3

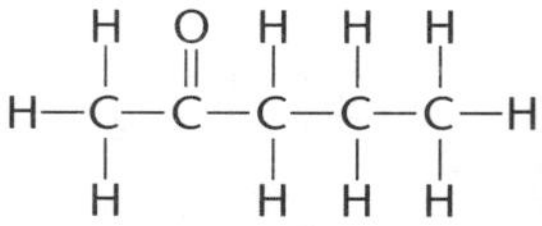

pentan-2-one
$CH_3COCH_2CH_2CH_3$

Uses of aldehydes and ketones

Propanone, $(CH_3)_2C{=}O$, is used in large quantities by industry as:

- an industrial solvent in paints, varnishes and nail-polish removers.

Methanal, $H_2C{=}O$ is used in large quantities in:

- preserving and embalming, as a germicide and insecticide
- for manufacturing plastic coatings such as *Bakelite*, *Formica* and melamine
- for manufacturing polymer adhesives such as those used to glue wood together.

Other aldehydes and ketones are mainly used as intermediates in the manufacture of plastics, dyes and pharmaceuticals, as solvents or as perfumes and flavouring agents.

Examples of aldehydes and ketones as flavours are:

- benzaldehyde, C_6H_5CHO (the flavour of fresh almonds)
- heptan-2-one, $C_5H_{11}COCH_3$ (smell of blue cheese).

Polarity of carbonyl compounds

Carbon and oxygen have different electronegativities, resulting in a polar C=O bond: carbonyl compounds have polar molecules.

The properties of aldehydes and ketones are dominated by the polar carbonyl group, C=O.

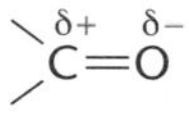

oxygen is more electronegative than carbon producing a dipole

Physical properties of carbonyl compounds

The polarity in propanone is such that it mixes with polar solvents such as water and also dissolves many organic compounds.

The low boiling point also makes it easy to remove by evaporation, a property exploited by its use in paints and varnishes.

The polarity of the carbonyl group is less than that of the hydroxyl group in alcohols. Thus aldehydes and ketones have weaker dipole-dipole interactions and lower boiling points than alcohols of comparable molecular mass.

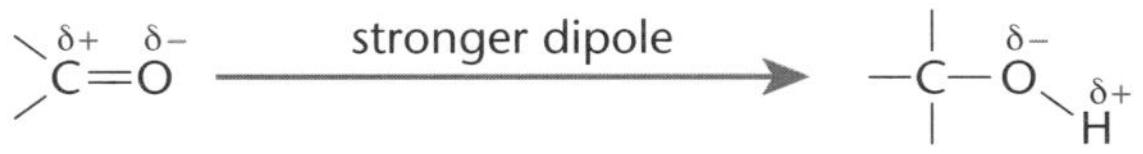

carbonyl group in aldehyde or ketone — hydroxyl group in alcohols

Nucleophilic addition to carbonyl compounds

AQA	M4	SALTERS	M5
EDEXCEL	M4, M5	WJEC	CH4
OCR	M4	NICCEA	M4
NUFFIELD	M4		

The electron-deficient carbon atom of the polar $C^{\delta+}{=}O^{\delta}$ bond attracts **nucleophiles**. This allows an **addition** reaction to take place across the C=O double bond of aldehydes and ketones. This is called **nucleophilic addition**.

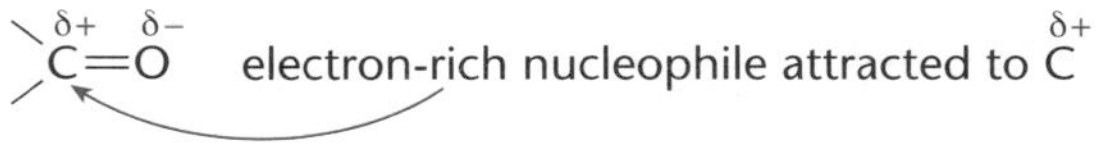

Nucleophilic addition of hydrogen cyanide, HCN

HCN is added across the C=O double bond.

In the presence of cyanide ions, CN^-, hydrogen cyanide, HCN, is added across the C=O bond in aldehydes and ketones.

aldehyde + HCN ⟶ hydroxynitrile

$CH_3CHO + HCN \longrightarrow CH_3CH(OH)CN$

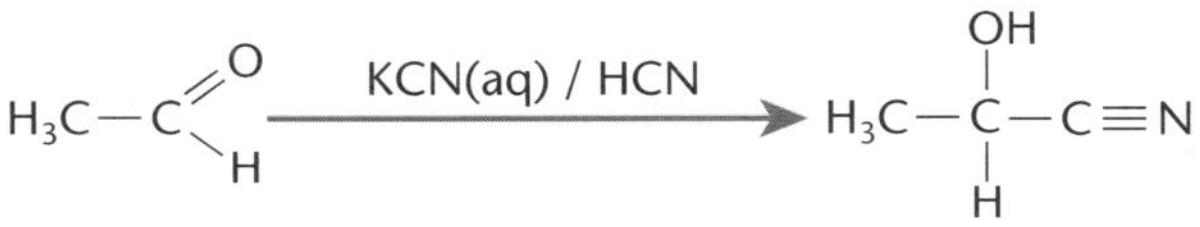

This hydroxynitrile has optical isomers.

Hydrogen cyanide is a very poisonous gas and it is usually generated in solution as H^+ and CN^- ions using:

- sodium cyanide, NaCN as a source of CN^-
- dilute sulphuric acid as a source of H^+.

Mechanism

The presence of cyanide ions is essential to provide the nucleophile for the first step of this mechanism.

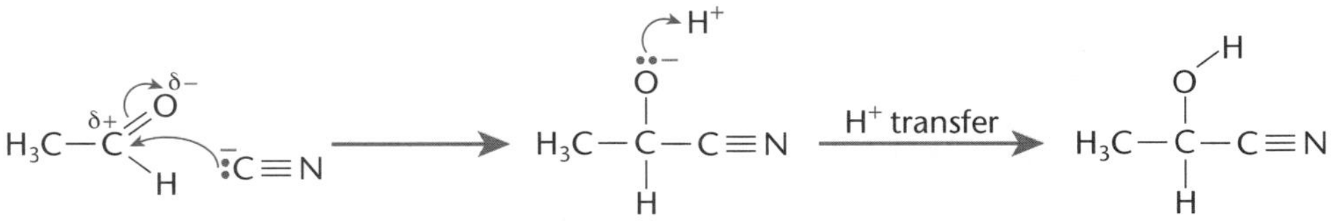

:CN^- nucleophile donates electron pair

'hydroxynitrile'

Increasing the carbon chain length

The nucleophilic addition of hydrogen cyanide is useful for increasing the length of a carbon chain. The nitrile product can then easily be reacted further in organic synthesis.

- Nitriles are easily ***hydrolysed*** by water in hot dilute acid to form a carboxylic acid.
- Nitriles are easily ***reduced*** by sodium in ethanol to form an amine.

The diagrams below show how 2-hydroxypropanenitrile, synthesised above, can be converted into a carboxylic acid and an amine.

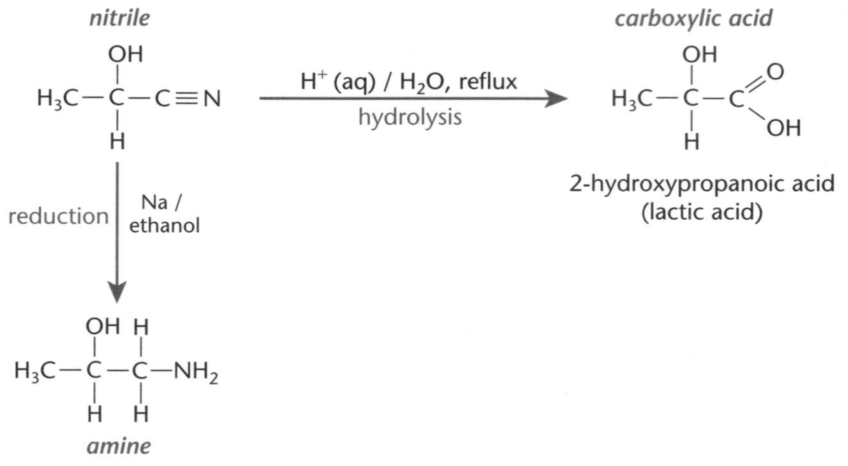

Testing for the carbonyl group

EDEXCEL	M4, M5	WJEC	CH4
OCR	M4	NICCEA	M4
NUFFIELD	M4		

Carbonyl compounds produce an orange-yellow crystalline solid with Brady's reagent.

The carbonyl group can also be detected using i.r. spectroscopy (see pp. 132–133).

The carbonyl group can be detected using **Brady's reagent** – a solution of 2,4-dinitrophenylhydrazine (2,4-DNPH) in dilute acid.

- With 2,4-DNPH, both aldehydes and ketones produce bright **orange-yellow crystals** which identifies the carbonyl group, C=O.

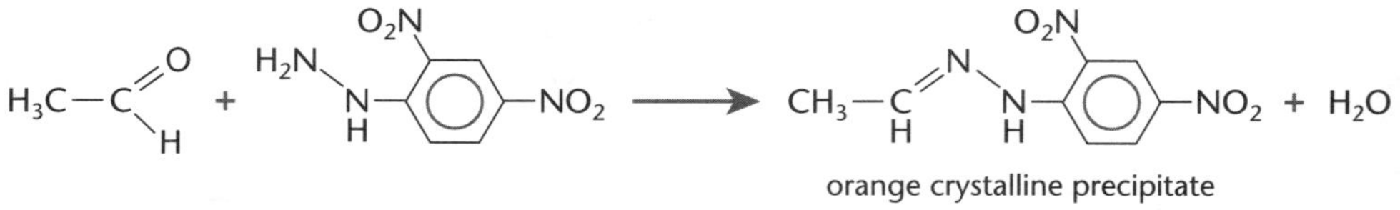

This is a **condensation reaction** – water is lost.

- The crystals have very sharp melting points which can be compared with known melting points from databases. Thus the actual carbonyl compound can be identified.

Testing for the aldehyde group

AQA	M4	NUFFIELD	M4
EDEXCEL	M4, M5	WJEC	CH4
OCR	M4	NICCEA	M4

Key points from AS

- **Oxidation of alcohols**
  Revise AS pages 128–129

An aldehyde can be distinguished from a ketone by using a combination of these two tests.

Both aldehyde and ketone produce a yellow-orange precipitate with 2,4-DNPH.

Only the aldehydes react in the three tests below.

You studied the oxidation of different types of alcohols during AS Chemistry. This is an extremely important reaction in organic chemistry, linking alcohols with aldehydes, ketones and carboxylic acids.

In terms of oxidation:

- aldehydes sit midway between primary alcohols and carboxylic acids.

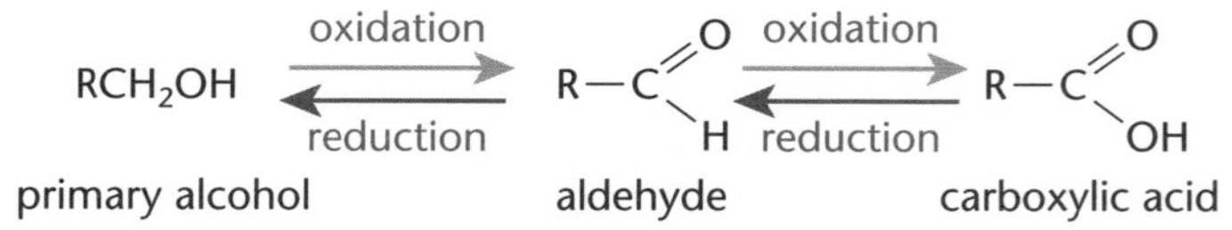

The oxidation of aldehydes provides the basis of chemical tests used to identify this functional group.

> **KEY POINT**
>
> In each test:
> - the aldehyde **reduces** the reagents used in the test, producing a visible colour change
> - the aldehyde is **oxidised** to a carboxylic acid.
> $RCHO + [O] \longrightarrow RCOOH$

Ketones cannot normally be oxidised and they **do not** react with Tollens' reagent, Fehling's solution or acidified dichromate(VI).

Heat aldehyde with Tollens' reagent

Tollens' reagent is a solution of silver nitrate in aqueous ammonia. The silver ions are reduced by an aldehyde producing a **silver mirror**.

> **KEY POINT**
>
> Tollens' reagent $\longrightarrow$ **silver mirror**
> $Ag^+(aq) + e^- \longrightarrow Ag(s)$

Heat aldehyde with Fehling's (or Benedict's) solution

Fehling's and Benedict's solutions contain Cu^{2+} ions dissolved in aqueous alkali. The Cu^{2+} ions are reduced to copper(I) ions, Cu^+, producing a **brick-red precipitate** of Cu_2O.

Key points from AS

- **Oxidation of primary alcohols**
  Revise AS pages 128–129

> **KEY POINT**
>
> Benedict's / Fehling's solution $\longrightarrow$ **brick-red precipitate** of Cu_2O
> $2Cu^{2+}(aq) + 2e^- + 2OH^-(aq) \longrightarrow Cu_2O(s) + H_2O(l)$

Acidified dichromate(VI) can also be used to oxidise alcohols.

Heat aldehyde with $H^+/Cr_2O_7^{2-}$

Concentrated sulphuric acid, H_2SO_4, is used as a source of H^+ ions and potassium dichromate(VI), $K_2Cr_2O_7$, as a source of $Cr_2O_7^{2-}$ ions. The **orange** $Cr_2O_7^{2-}$ ions are reduced to **green** chromium(III) ions, Cr^{3+}.

Reduction of carbonyl compounds

AQA	M4	SALTERS	M5
EDEXCEL	M4	WJEC	CH4
OCR	M4	NICCEA	M4
NUFFIELD	M4		

Key point from AS

- **Reduction of carbonyl compounds**
  Revise AS page 129

Aldehydes and ketones can be reduced to alcohols using a reducing agent containing the hydride ion, H^-.

Suitable reducing agents are:
- sodium tetrahydridoborate(III) (*sodium borohydride*), $NaBH_4$, in water (reflux)
- lithium tetrahydridoaluminate(III) (*lithium aluminium hydride*), $LiAlH_4$, in dry ether (at room temperature).

Aldehydes are reduced to primary alcohols.

For balanced equations, the reducing agent can be shown simply as [H].

$$H_3C-C(=O)(H) + 2[H] \longrightarrow H_3C-CH_2OH$$

aldehyde   primary alcohol

$NaBH_4$ and $LiAlH_4$ both reduce the C=O double bond in aldehydes and ketones.

They do **not** reduce the C=C bond in alkenes.

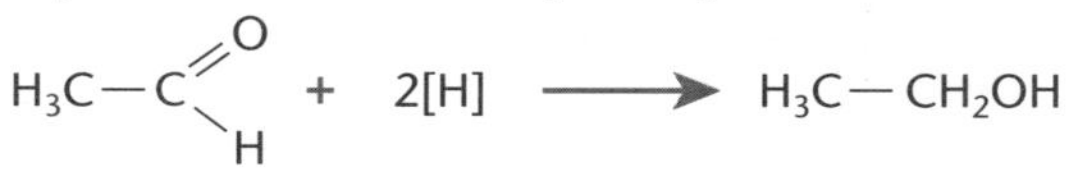

Ketones are reduced to secondary alcohols.

$$\begin{matrix} H_3C \\ C_2H_5 \end{matrix}C=O + 2[H] \longrightarrow \begin{matrix} H_3C \\ C_2H_5 \end{matrix}CHOH$$

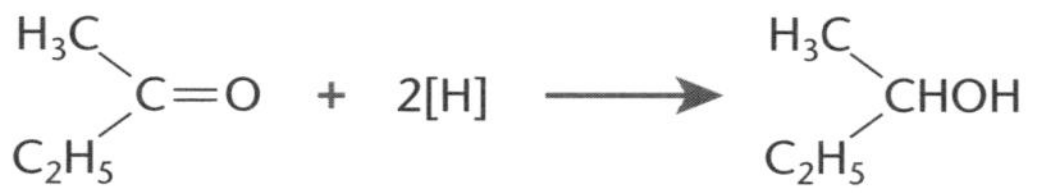

ketone   secondary alcohol

Mechanism (AQA only)

The reaction of carbonyl compounds with hydrogen cyanide (see p. 95) is also an example of nucleophilic addition.

The reduction of aldehydes and ketones to alcohols using $NaBH_4$ or $LiAlH_4$ is an example of **nucleophilic addition**. The reducing agent can be considered to release hydride ions, H^-.

$$NaBH_4 \longrightarrow H^- + NaBH_3^+$$

The hydride ion acts as a nucleophile in the first stage of the reaction.

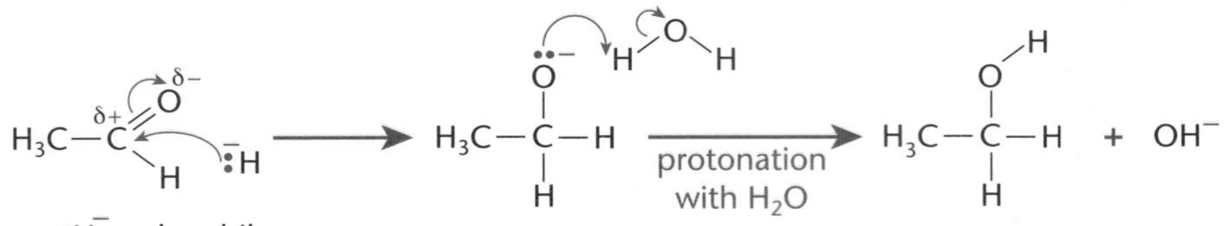

Test for the presence of the methyl carbonyl group

EDEXCEL M4, M5 WJEC CH4

The presence of a methyl carbonyl group can be detected by heating a compound with iodine, I_2 in aqueous sodium hydroxide, $OH^-(aq)$.

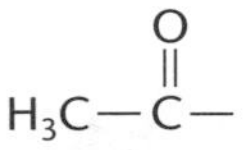

methyl carbonyl

Methyl carbonyl compounds produce pale-yellow crystals with I_2/OH^-.

- Methyl carbonyls produce **pale-yellow crystals** of triiodomethane (*iodoform*), CHI_3, with an antiseptic smell.
- $CH_3CHO + 3I_2 + 4OH^- \longrightarrow CHI_3 + HCOO^- + 3H_2O + 3I^-$

Progress check

1 (a) Draw the structure of:
(i) 3-methylpentan-2-one
(ii) 3-ethyl-2-methylpentanal.

2 **Three** isomers of C_4H_8O are carbonyl compounds.
(a) (i) Show the formula for each isomer.
(ii) Classify each isomer as an *aldehyde* or a *ketone.*
(iii) Name each isomer.
(b) Write the structural formulae of the **three** alcohols that could be formed by reduction of these isomers.

1 (a) (i)

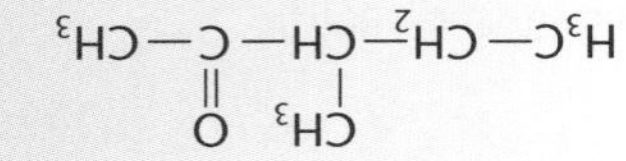

(ii)

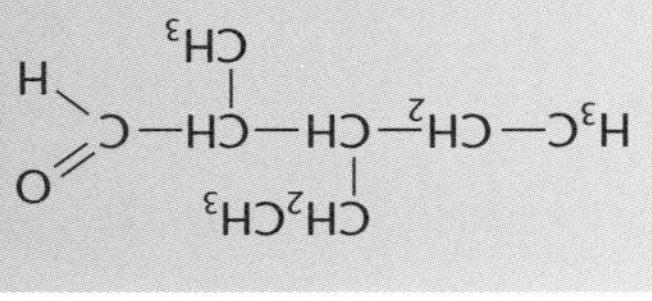

2 (a) (i)

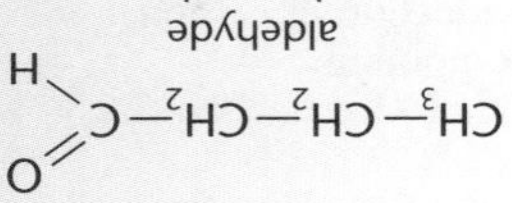

aldehyde
butanal

(ii)

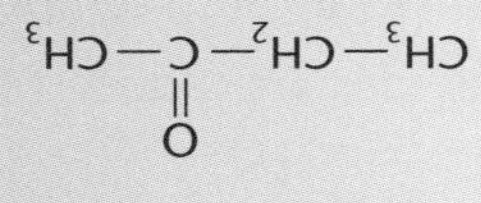

ketone
butanone

(iii)

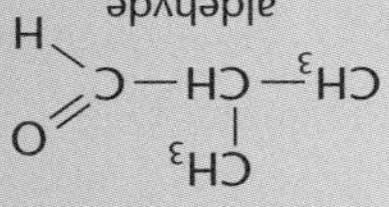

aldehyde
methylpropanal

(b) $CH_3CH_2CH_2CH_2OH$; $CH_3CH_2CHOHCH_3$; $(CH_3)_2CHCH_2OH$

5.3 Carboxylic acids

LEARNING SUMMARY

After studying this section you should be able to:

- *understand the polarity and physical properties of carboxylic acids*
- *describe acid reactions of carboxylic acids to form salts*
- *describe the esterification of carboxylic acids with alcohols*

Carboxylic acids General formula: $C_nH_{2n+1}COOH$ RCOOH

AQA	M4	SALTERS	M4
EDEXCEL	M4	WJEC	CH4
OCR	M4	NICCEA	M4
NUFFIELD	M4		

The combination of carbonyl and hydroxyl groups in the carboxyl group modifies the chemistry of both groups.

Carboxylic acids have their own set of reactions and react differently from carbonyl compounds and alcohols.

The carboxyl group

The functional group in carboxylic acids is the **carboxyl** group, COOH. Although this combines both the **carbon**yl group and a hydrox**yl** group its properties are very different from either.

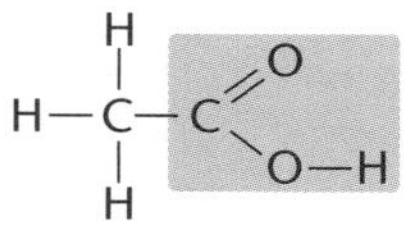

carboxyl group

Naming of carboxylic acids

Numbering starts from the carbon atom of the carboxyl group.

4-methylpentanoic acid
$(CH_3)_2CH_2CH_2CH_2COOH$

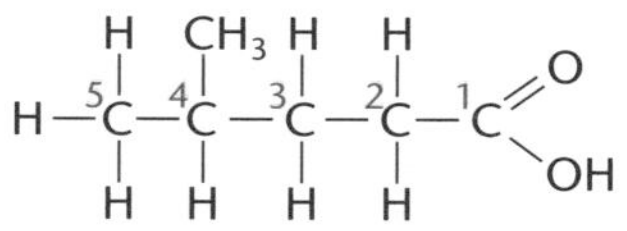

Natural carboxylic acids

Carboxylic acids are found commonly in nature. Their acidity is comparatively weak and their presence in food often gives a sour taste. Some examples of natural carboxylic acids are shown below.

structure	*name*	*natural source*
HCOOH	methanoic acid (*formic acid*)	ants, stinging nettles
CH_3COOH	ethanoic acid (*acetic acid*)	vinegar
COOH–COOH	ethanedioic acid (*oxalic acid*)	rhubarb
$H_3C-CH(OH)-COOH$	2-hydroxypropanoic acid (*lactic acid*)	sour milk
$HO-C(COOH)(CH_2-COOH)_2$	2-hydroxypropane-1,2,3-tricarboxylic acid (*citric acid*)	oranges, lemons

Polarity

The carboxyl group is a combination of two polar groups: the hydroxyl –OH, **and** carbonyl C=O groups. This makes a carboxylic acid molecule more polar than a molecule of an alcohol or a carbonyl compound.

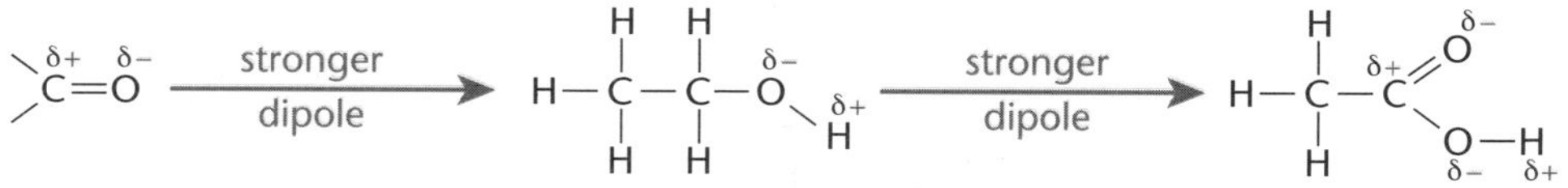

carbonyl group in aldehyde or ketone — hydroxyl group in alcohols — carboxyl group in carboxylic acids

The properties of carboxylic acids are dominated by the carboxyl group, COOH.

Physical properties of carboxylic acids

The COOH group dominates the physical properties of short-chain carboxylic acids. Hydrogen bonding takes place between carboxylic acid molecules, resulting in:

- higher melting and boiling points than alkanes of comparable M_r
- solubility in water.

The solubility of alcohols in water decreases with increasing carbon chain length as the non-polar contribution to the molecule becomes more important.

Carboxylic acids as 'acids'

AQA	M4	SALTERS	M4
EDEXCEL	M4	WJEC	CH4
OCR	M4	NICCEA	M4
NUFFIELD	M4		

Carboxylic acids are the 'organic acids'.

For more details of the dissociation of weak acids, see p. 36.

Carboxylic acids are only weak acids because they only partially dissociate in water.

$$CH_3COOH \rightleftharpoons CH_3COO^- + H^+$$

- Only 1 molecule in about 100 actually dissociates.
- Only a small proportion of the potential H^+ ions is released.

Carboxylates: salts of carboxylic acids

Carboxylic acid salts, 'carboxylates', are formed by neutralisation of a carboxylic acid by an alkali. In the example below, ethanoic acid produces ethanoate ions.

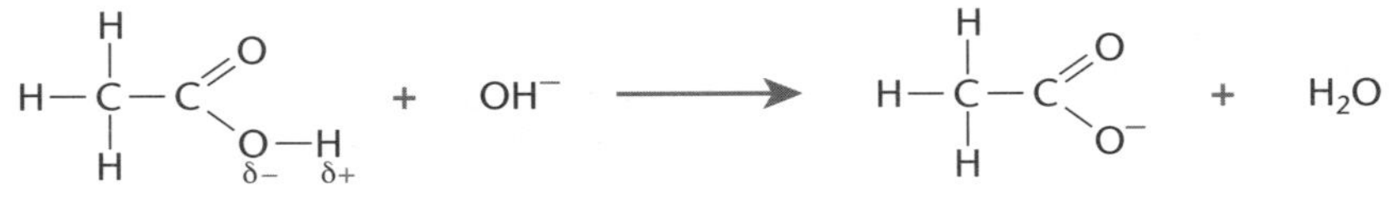

carboxylic acids exist in acidic conditions

carboxylates exist in alkaline conditions

On evaporation of water, a carboxylate salt crystallises out as an ionic compound. With aqueous sodium hydroxide as the alkali, sodium ethanoate, $CH_3COO^-Na^+$, is produced as the ionic salt.

Acid reactions of carboxylic acids

Key points from AS

- **Typical reactions of an acid**
  Revise AS page 110

Carboxylic acids react by the usual 'acid reactions' producing carboxylate salts.

Carboxylic acids take part in typical acid reactions. Note in the examples below that each salt formed is a carboxylate.

- They are **neutralised by alkalis**, forming a salt and water only.
  $CH_3COOH + NaOH \longrightarrow CH_3COONa + H_2O$
- They **react with carbonates** forming a salt, carbon dioxide and water.
  $2CH_3COOH + CaCO_3 \longrightarrow (CH_3COO)_2Ca + CO_2 + H_2O$
- They **react with** reactive **metals** forming a salt and hydrogen.
  $2CH_3COOH + Mg \longrightarrow (CH_3COO)_2Mg + H_2$

KEY POINT

Carboxylic acids are the only common organic group able to release carbon dioxide gas from carbonates. This provides a useful test to show the presence of a carboxyl group.

Properties of carboxylates

Carboxylates such as sodium ethanoate, $CH_3COO^-Na^+$, are ionic compounds. They have typical properties of an ionic compound.

- They are solids at room temperature with high melting and boiling points.
- They have a giant ionic lattice structure.
- They dissolve in water, totally dissociating into ions.

Reduction of carboxylic acids

EDEXCEL M4 | NICCEA M4 | WJEC CH4

Just as primary alcohols can be oxidised to carboxylic acids, the reverse process can take place using a strong reducing agent such as lithium tetrahydridoalumuninate(III) (*lithium aluminium hydride*), $LiAlH_4$ in dry ether. (See also: Reduction of carbonyl compounds, page 97.)

$$RCOOH + 4[H] \longrightarrow RCH_2OH + H_2O$$

Esterification

AQA M4 | SALTERS M4 | EDEXCEL M4 | WJEC CH4 | OCR M4 | NICCEA M4 | NUFFIELD M4

Esterification is the formation of an ester by reaction of a **carboxylic acid** with an **alcohol** in the presence of an **acid catalyst** (e.g. concentrated sulphuric acid).

carboxylic acid + alcohol ⟶ ester + water

The esterification of ethanoic acid by methanol is shown below:

Esters are formed by reaction of a carboxylic acid with an alcohol.

$$CH_3COOH + CH_3OH \longrightarrow CH_3COOCH_3 + H_2O$$

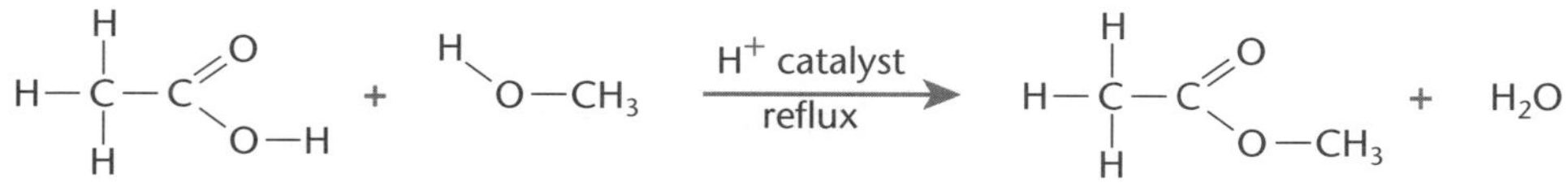

Conditions

- An acid catalyst (a few drops of conc. H_2SO_4) and reflux.
- The yield is usually poor due to incomplete reaction.

Progress check

1 Write down the structural formula of:
(a) propanoic acid
(b) the propanoate ion.

2 Explain why a carboxylic acid has a higher boiling point than the corresponding alcohol.

1 (a) CH_3CH_2COOH
(b) $CH_3CH_2COO^-$

2 Carboxylic acid has both polar carbonyl and hydroxyl groups. Alcohols have hydroxyl group only. Therefore, a carboxylic acid has greater intermolecular forces.

5.4 Esters

LEARNING SUMMARY

After studying this section you should be able to:

- *understand the physical properties of esters*
- *describe the acid and base hydrolysis of esters*
- *describe the hydrolysis of fats and oils in soap making*

Esters

General formula: $C_nH_{2n+1}COOC_mH_{2m+1}$ RCOOR'

AQA	M4	SALTERS	M4
EDEXCEL	M4	WJEC	CH4
OCR	M4	NICCEA	M4
NUFFIELD	M4		

The functional group in esters is the COOR' group, comprises an alkyl group in place of the acidic proton of a carboxylic acid.

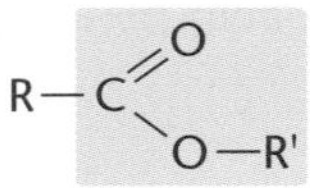

Naming of esters

The name of an ester is based on the carboxylic acid from which the ester is derived. The ester below, methyl butanoate, is derived from butanoic acid C_3H_7COOH with a methyl group in place of the acidic proton.

butanoate from butanoic acid C_3H_7COOH

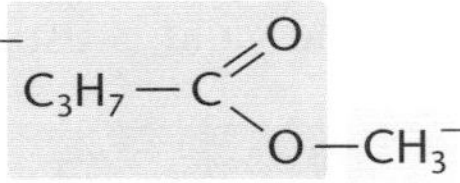

methyl from methanol CH_3OH

methyl butanoate

In the name of **methyl butanoate**,

- the alkyl group comes first as the ......*yl*: **methyl**
- the carboxylic acid part comes second as the ...*oate*: **butanoate**

Natural esters

Esters are found commonly in nature as fats and oils. They often have pleasant smells and contribute to the flavouring of many foods.

Some examples of natural esters are shown below.

Esters are common organic compounds present in fats and oils. They are also used in food flavourings and as perfumes.

structure	*name*	*source*
$HCOOCH_3$	methyl methanoate	raspberries
$C_3H_7COOC_4H_9$	butyl butanoate	pineapple
$CH_3COOCH_2CH_2CH(CH_3)_2$	3-methylbutyl ethanoate	pears

Physical properties of esters

Unlike carboxylic acids, esters are neutral. They are less polar than carboxylic acids and the absence of an –OH group means that esters cannot form hydrogen bonds and are generally insoluble in water.

Hydrolysis of esters

AQA	M4	SALTERS	M4
EDEXCEL	M4	WJEC	CH4
OCR	M4	NICCEA	M4
NUFFIELD	M4		

The hydrolysis of an ester is the reverse reaction to esterification (see page 101):

$$\text{ester} + \text{water} \underset{\textit{esterification}}{\overset{\textit{hydrolysis}}{\rightleftharpoons}} \text{carboxylic acid} + \text{alcohol}$$

Note that this reaction is reversible. The direction of reaction can be controlled by the reagents and reaction conditions used.

> **KEY POINT**
> Hydrolysis is the breaking down of a compound using **water** as the reagent.

Esterification **produces** water.

Hydrolysis **reacts** with water.

Hydrolysis takes place by refluxing the ester with dilute aqueous acid or alkali.

- Acid hydrolysis ⟶ alcohol + carboxylic acid.
- Alkaline hydrolysis ⟶ alcohol + carboxylate.

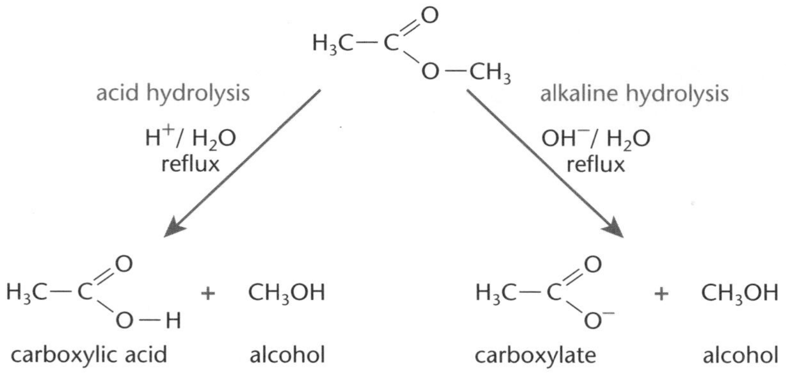

Hydrolysis of fats and oils

AQA	M4	SALTERS	M5
OCR	M4	NICCEA	M4

Fats are *triglyceryl esters* of fatty acids (long chain carboxylic acids) and propane-1,2,3-triol (*glycerol*).

The hydrolysis of fats and oils is an important reaction used in soap production. Acid hydrolysis of each molecule of the fat produces:

- **three** fatty acid molecules, 3RCOOH and
- **one** molecule of glycerol (a triol), $HOCH_2CHOHCH_2OH$.

Alkaline hydrolysis produces the carboxylate of the fatty acid.

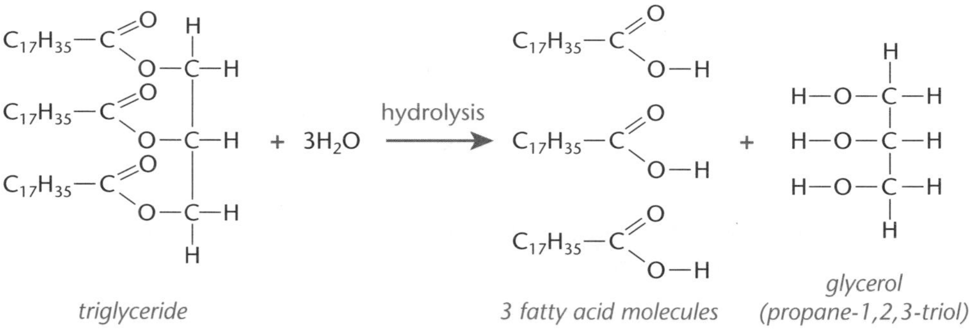

Soaps are carboxylate salts, prepared by the alkaline hydrolysis of fats and oils. This process is called **saponification** from the Latin *sapo*: soap.

Alkaline hydrolysis with aqueous sodium hydroxide of the triglyceride above forms the sodium carboxylate salt $C_{17}H_{35}COO^-Na^+$.

Progress check

1 How could you prepare methyl propanoate from a named carboxylic acid and alcohol?

2 (a) Explain what is meant by the term *hydrolysis*.
(b) Write down the products of the hydrolysis of ethyl butanoate in acidic and alkaline conditions.

1 Reflux propanoic acid with methanol in the presence of a few drops of concentrated sulphuric acid as catalyst.
2 (a) Breaking down a molecule using water.
(b) Acid hydrolysis: ethanol and butanoic acid.
Alkaline hydrolysis: ethanol and butanoate ions.

5.5 Acylation

After studying this section you should be able to:

- *describe the formation of acyl chlorides from carboxylic acids*
- *describe nucleophilic addition-elimination reactions of acyl chlorides*
- *know the advantages of ethanoic anhydride for industrial acylations*

LEARNING SUMMARY

Acyl chlorides

General formula: $C_nH_{2n+1}COCl$ RCOCl

AQA	M4	SALTERS	M4
EDEXCEL	M4	WJEC	CH4
NUFFIELD	M4	NICCEA	M4

The functional group in acyl chlorides is the COCl group,

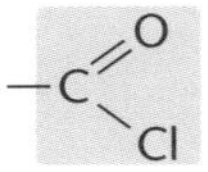

Acyl chlorides are named from the parent carboxylic acid.

The examples below show that the suffix '*-oic acid*' is changed to '*-oyl chloride*' in the corresponding acyl chloride.

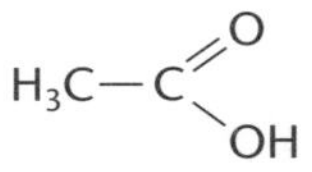
ethanoic acid

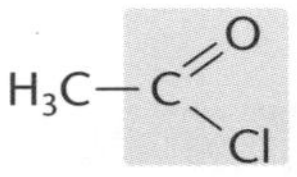
ethanoyl chloride

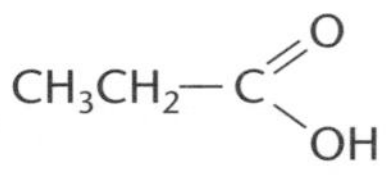
propanoic acid

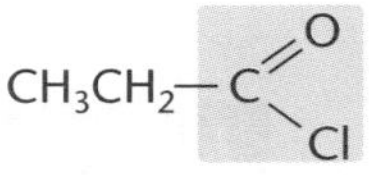
propanoyl chloride

Properties

Acyl chlorides are the most reactive organic compounds that are commonly used. Unlike most other organic functional groups, acyl chlorides do **not** occur naturally because of their high reactivity with water.

Acyl chlorides are the only common organic compounds that react violently with water.

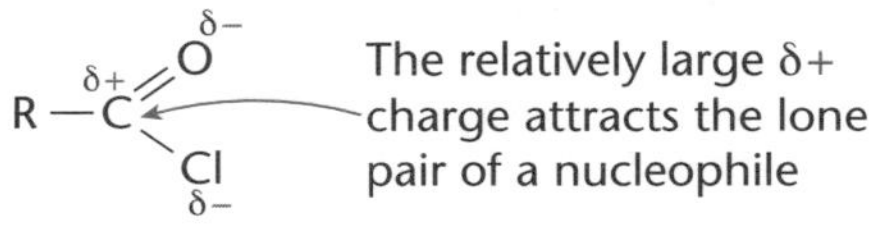
The relatively large δ+ charge attracts the lone pair of a nucleophile

The electron-withdrawing effects of both the **chlorine** atom and carbonyl **oxygen** atom produces a relatively large δ+ charge on the carbonyl carbon atom. Nucleophiles are strongly attracted to the electron-deficient carbon atom, increasing the reactivity.

Acyl chlorides in organic synthesis

AQA	M4	WJEC	CH4
EDEXCEL	M4	NICCEA	M4
NUFFIELD	M4		

The high reactivity of acyl chlorides compared with carboxylic acids makes them particularly useful in the organic synthesis of related compounds.

Advantages of acyl chlorides over carboxylic acids.

- A good yield of product – reactions go to completion.
- Reactions often occur quickly and at lower temperatures.

Using acyl chlorides in synthesis

The high reactivity of an acyl chloride means that it is usually prepared from the parent carboxylic acid '*in situ*', i.e. when it is needed.

Acyl chlorides are prepared by reacting a carboxylic acid with:

- phosphorus pentachloride

$$RCOOH + PCl_5 \longrightarrow RCOCl + POCl_3 + HCl$$

- or sulphur dichloride oxide, $SOCl_2$

$$RCOOH + SOCl_2 \longrightarrow RCOCl + SO_2 + HCl$$

- The required nucleophile is then added to the acyl chloride.

With acyl chlorides, anhydrous conditions are **essential** – acyl chlorides react with water.

Reactions of acyl chlorides with nucleophiles

AQA	M4	SALTERS	M4
EDEXCEL	M4, M6	WJEC	CH4
NUFFIELD	M4	NICCEA	M4

Nucleophiles of the type H–Y**:** react readily with acyl chlorides.

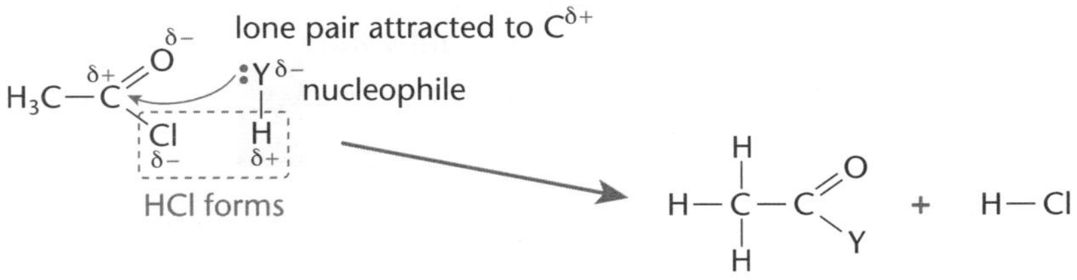

This is an **addition-elimination** reaction involving:

- addition of HY across the C=O double bond followed by
- elimination of HCl.

Addition-elimination reactions of acyl chlorides

Acyl chlorides can be reacted with different nucleophiles to produce a range of related functional groups. In each addition-elimination reaction, hydrogen chloride is eliminated as the second product.

The reaction scheme below shows the reactions of an acyl chloride RCOCl with the nucleophiles water, H_2O, methanol, CH_3OH, ammonia, NH_3 and methylamine, CH_3NH_2.

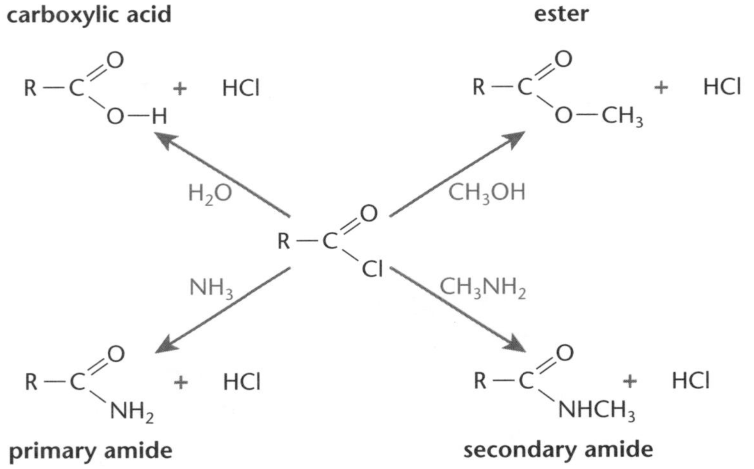

The high reactivity of an acyl chloride means that these reactions take place at room temperature.

Ammonia and methylamine are bases – they react with the HCl eliminated:

$$NH_3 + HCl \longrightarrow NH_4^+Cl^-$$
$$CH_3NH_2 + HCl \longrightarrow CH_3NH_3^+Cl^-$$

Mechanism (AQA only)

The mechanism below shows the **nucleophilic addition-elimination** reaction of ethanoyl chloride with methanol to form an ester. Each nucleophile in the reaction scheme above will react in a similar manner.

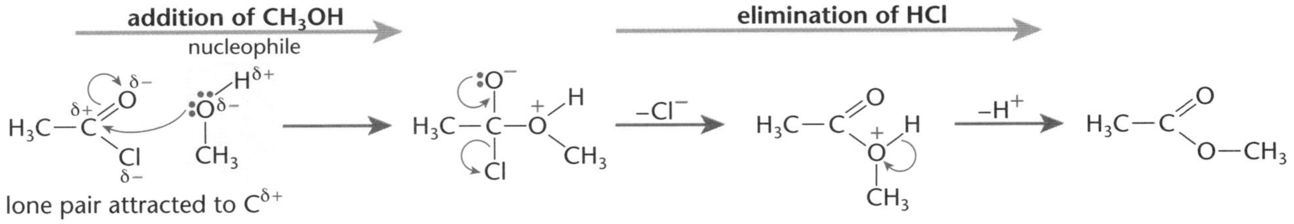

Acid anhydrides

AQA M4 WJEC CH4

Acyl chlorides are ideal for small-scale preparations in the laboratory. However, they are too expensive and too reactive for large-scale preparations for which they are replaced by acid anhydrides.

The equation below shows the reaction between ethanoic anhydride and methanol.

Notice the molecular structure of ethanoic anhydride – two ethanoic acid molecules have been condensed together with loss of H_2O. Hence the 'anhydride' from ethanoic acid.

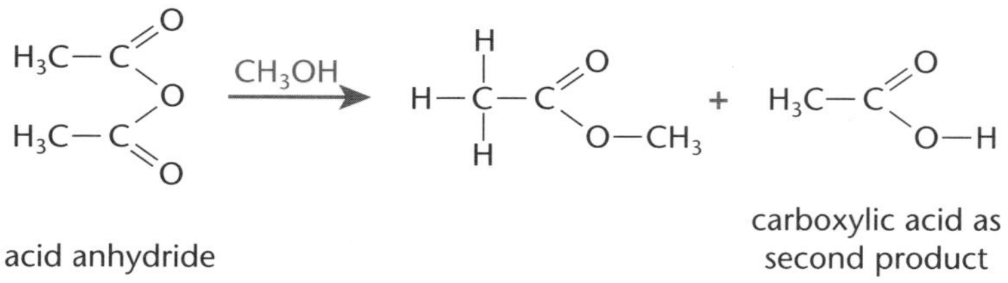

- The reaction is slower and easier to control on a large scale.
- This is the same essential reaction as with an acyl chloride but RCOOH forms instead of HCl as the second product.

Synthesis of aspirin

Aspirin can be synthesised by acylation of 2-hydroxybenzenecarboxylic acid (*salicylic acid*) using ethanoic anhydride.

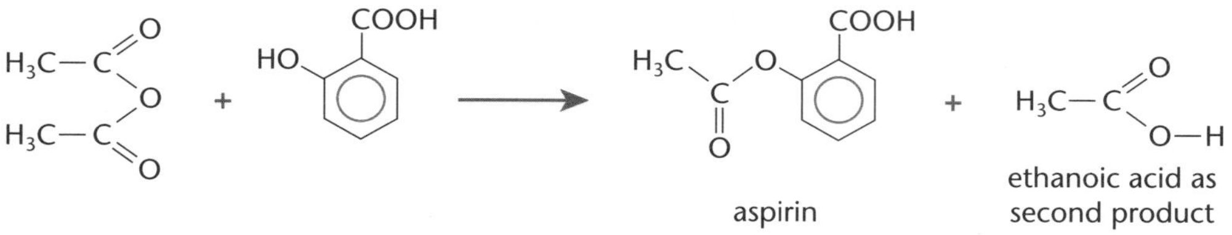

Progress check

1 Write down the structural formula of the organic product formed from the reaction of propanoyl chloride with:
(a) water
(b) ammonia.

2 What could you react together to make N-ethylbutanamide, $CH_3CH_2CH_2CONHCH_2CH_3$?

3 Write an equation for the preparation of aspirin using an acyl chloride.

1 (a) CH_3CH_2COOH
(b) $CH_3CH_2CONH_2$
2 $CH_3CH_2CH_2COCl$ and $CH_3CH_2NH_2$
3

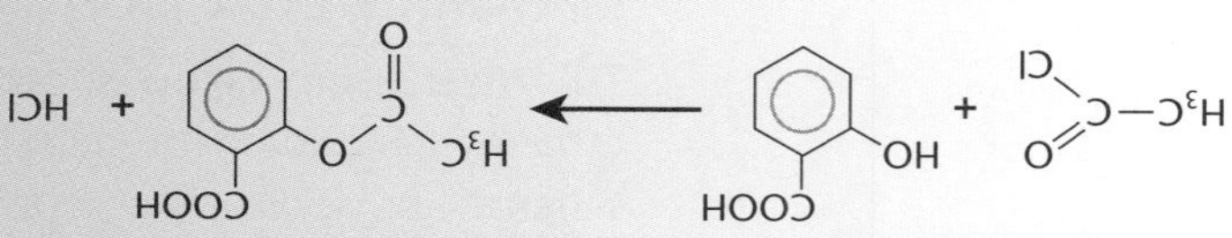

Sample question and model answer

(a) Outline the reaction of propanone with the following reagents. Give the equation for the reaction, the conditions, and the name of the organic product.

(i) Hydrogen cyanide

Equation:

$(CH_3)_2CO + HCN \longrightarrow (CH_3)_2C(OH)CN$ ✓

Conditions:

NaCN/dilute acid, heat ✓

Name of product:

ethanal hydroxynitrile ✓ [3]

In many questions, such as this, based on organic chemistry, you will be rewarded if you thoroughly learn the reactions, reagents and conditions.

(ii) Sodium tetrahydridoborate(III) (sodium borohydride).

Equation (you may represent $NaBH_4$ as [H]):

$(CH_3)_2CO + 2[H] \longrightarrow (CH_3)_2CHOH$ ✓

Conditions:

H_2O/reflux ✓

Name of product:

propan-2-ol ✓ [3]

(b) (i) Give a mechanism for the reaction in (a)(i).

Notice the care taken with the arrows.

Notice that the lone pair of the nucleophile is always shown.

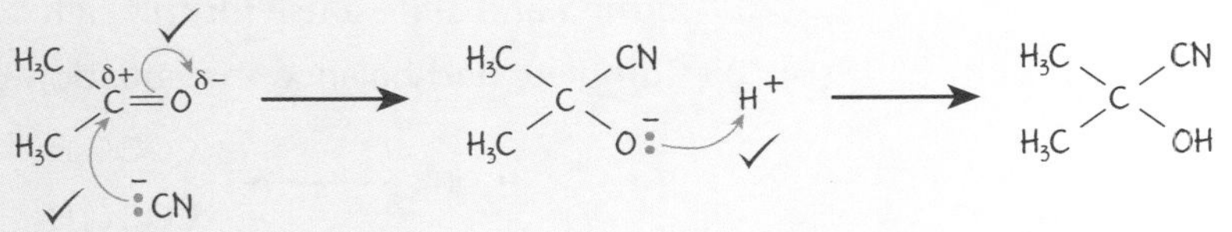

[3]

(ii) What type of mechanism is this?

nucleophilic addition ✓ [1]

'Nucleophilic' must be included for the mark.

(iii) What feature of the carbonyl group makes this type of mechanism possible? Explain how this feature arises.

dipole across C=O resulting in an electron-deficient carbon atom which can attract a nucleophile; ✓

oxygen is more electronegative than carbon. ✓ [2]

(iv) Explain briefly, by reference to its structure, why ethene would not react with HCN in a similar way.

The C=C double bond is between two identical atoms. ∴ there is no dipole and no electron-deficient carbon atom to attract a nucleophile. ✓ [1]

This is more difficult but tested often in exams.

Alkenes react by **electrophilic** addition.

Carbonyl compounds react by **nucleophilic** addition.

[Total: 13]

Edexcel Module Test 4 Q2 June 1998

Practice examination questions

1

(a) Two isomeric compounds, **A** and **B**, each have a relative molecular mass of 58 and the following percentage composition by mass: C, 62.1%; H, 10.3%; O, 27.6%. Calculate the molecular formula of **A** and of **B**. [2]

(b) Draw the structural formula of **A** and of **B**. [2]

(c) Name a reagent that reacts both with **A** and with **B**. State the observation. [2]

(d) Name a reagent that reacts with only one of **A** and **B**. Identify which isomer reacts and state the observation. [2]

[Total: 8]

Cambridge Chains and Rings Q1 Nov 1999

2

(a) Compound **G**, shown below, is a tri-ester.

$$CH_3(CH_2)_{16}COO-CH_2$$
$$CH_3(CH_2)_{16}COO-CH_2$$
$$CH_3(CH_2)_{16}COO-CH_2$$

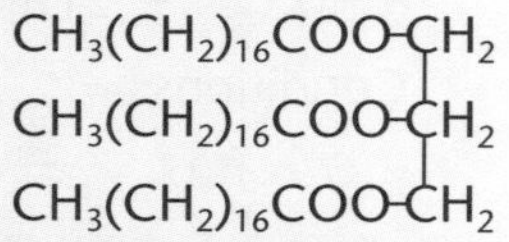

(i) Deduce the physical state of **G** at room temperature.

(ii) When completely hydrolysed by heating with aqueous sodium hydroxide, **G** forms an alcohol and the sodium salt of a carboxylic acid. Give the structural formula of the alcohol formed, write a formula for the sodium salt formed and state a use for this salt. [4]

(b) (i) Complete and balance the equation below for the formation of di-ester **H**.

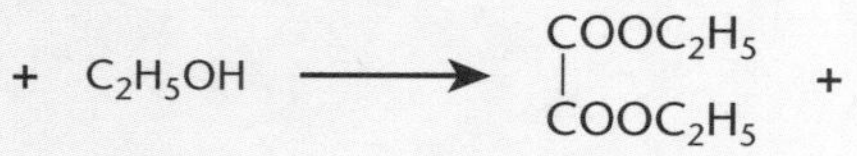

$$+ \ C_2H_5OH \ \longrightarrow \ \begin{matrix} COOC_2H_5 \\ | \\ COOC_2H_5 \end{matrix} \ +$$
$$\textbf{H}$$

(ii) Identify a substance which could catalyse this reaction to form **H**.

(iii) Draw a structural isomer of **H**, also a di-ester, which is formed by the reaction of ethane-1,2-diol with one other compound. [5]

[Total: 9]

Assessment and Qualification Alliance Kinetics and Organic Chemistry Q5 March 1998

3

Consider the reaction scheme shown below.

$$\begin{matrix} CH_3 \\ | \\ H-C=O \end{matrix} \xrightarrow{\text{step 1}} \begin{matrix} CH_3 \\ | \\ H-C-OH \\ | \\ CN \end{matrix} \xrightarrow{\text{step 2}} \begin{matrix} CH_3 \\ | \\ H-C-OH \\ | \\ COOH \\ \textbf{B} \end{matrix} \xrightarrow{H^+/Cr_2O_7^{2-}} \text{compound } \textbf{C}$$

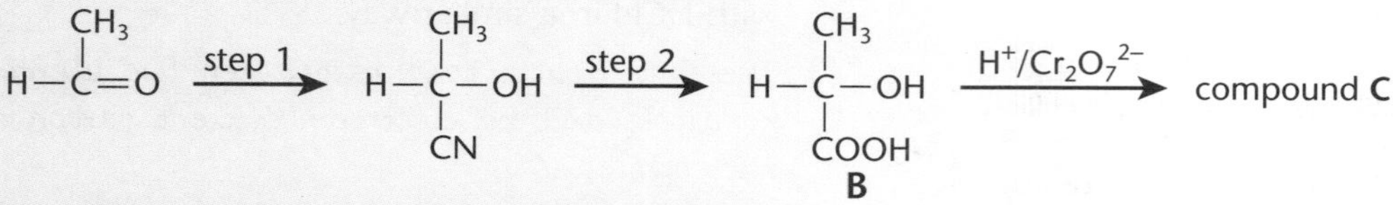

(a) State the reagent(s) that could be used for:
   (i) step 1 (ii) step 2. [2]

(b) Compound **B** has optical isomers. What structural feature in compound **B** results in optical isomerism?

(c) (i) Draw the structure of compound **C**. [1]

   (ii) Compound **C** can be reduced to propane-1,2-diol. Using [H] to represent the reducing agent, construct a balanced equation for this reduction. [3]

[Total: 6]

Cambridge Chains and Rings Q5 June 2000

Chapter 6

Aromatics, amines, amino acids and polymers

The following topics are covered in this chapter:

- *Arenes*
- *Reactions of arenes*
- *Phenols*
- *Amines*
- *Amino acids*
- *Polymers*

6.1 Arenes

LEARNING SUMMARY

After studying this section you should be able to:

- *apply rules for naming simple aromatic compounds*
- *understand the delocalised model of benzene*
- *explain the resistance to addition of benzene compared with alkenes*

Aromatic organic compounds

AQA	M4	SALTERS	M5
EDEXCEL	M5	WJEC	CH4
OCR	M4	NICCEA	M5
NUFFIELD	M4		

Organic compounds with pleasant smells were originally classified as aromatic compounds. Many of these contain a benzene ring in their structure and nowadays an aromatic compound is one structurally derived from benzene, C_6H_6.

The diagrams below show different representations of a benzene molecule. It is usual practice to omit the carbon and hydrogen labels.

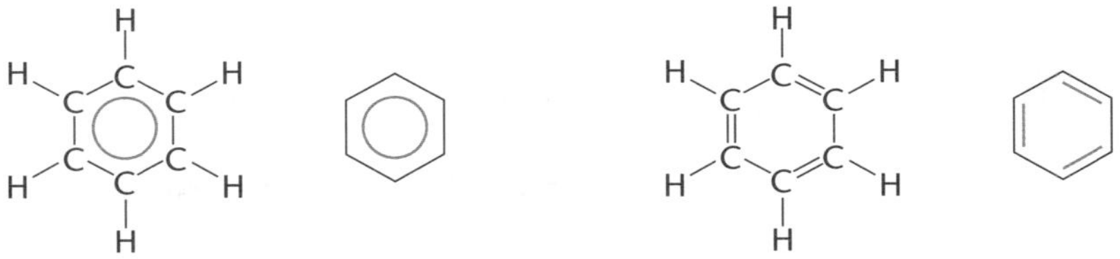

Arenes

An **arene** is an aromatic hydrocarbon containing a benzene ring. Benzene is the simplest arene and substituted arenes have alkyl groups attached to the benzene ring. Examples of arenes are shown below.

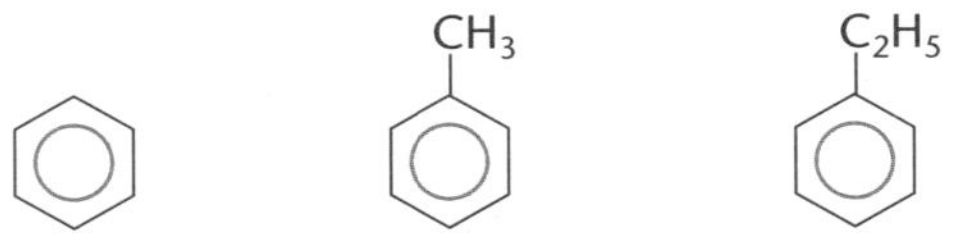

benzene methylbenzene ethylbenzene

Functional groups

Arenes can have a functional group next to an **aryl** group (a group containing a benzene ring):

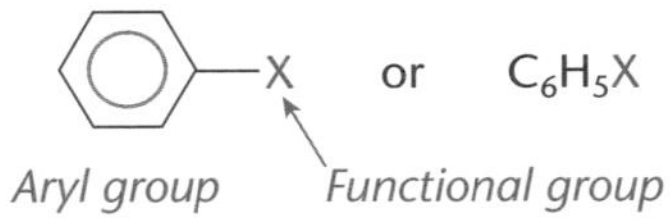

An aryl group contains a benzene ring.

- An aryl group is often represented simply as Ar—.
- The simplest aryl group is the **phenyl** group, C_6H_5, derived from benzene, C_6H_6.

Naming of aromatic organic compounds

The benzene ring of an aromatic compound is numbered from the carbon atom attached to a functional group or alkyl side-chain.
The names can be derived in two ways.

- Some compounds (e.g. hydrocarbons, chloroarenes and nitroarenes) are regarded as substituted benzene rings.

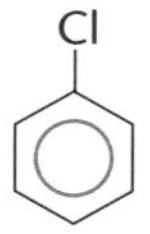

chlorobenzene

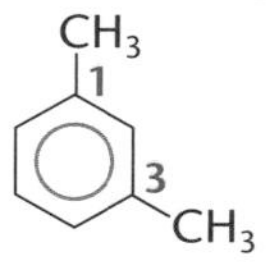

1,3-dimethylbenzene

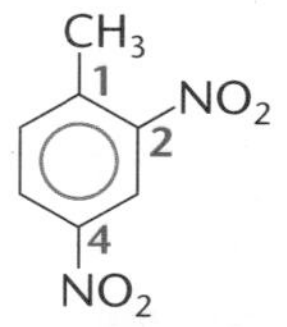

2,4-dinitromethylbenzene

You should be able to suggest names for simple arenes.

- Other compounds (e.g. phenols, and amines) are considered as phenyl compounds of a functional group.

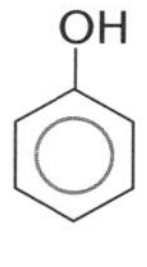

phenol
not hydroxybenzene

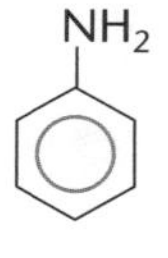

phenylamine
not aminobenzene

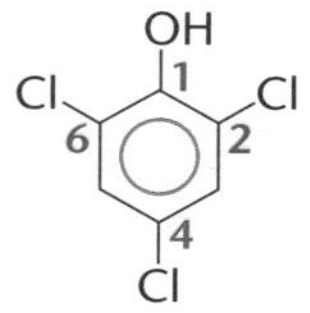

2,4,6-trichlorophenol

The stability of benzene

AQA	M4	SALTERS	M5
EDEXCEL	M5	WJEC	CH4
OCR	M4	NICCEA	M5
NUFFIELD	M4		

Two structures are used to represent benzene:

- the **Kekulé** structure developed between 1865 and 1872
- the modern **delocalised** structure developed in the 1930s.

The Kekulé structure of benzene

The Kekulé model of a benzene molecule has alternate double and single bonds making up the ring.

The original Kekulé structure showed benzene as a hexagonal molecule with alternate double and single bonds. Each carbon atom is attached to one hydrogen atom. This model was later modified to one with two isomers, rapidly interconverting into one another.

The Kekulé structure of benzene

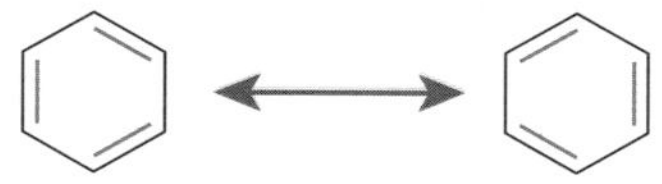

The chemical name for the Kekulé structure of benzene is **cyclohexa-1,3,5-triene**, after the positions of the double bonds in the ring.

The delocalised structure of benzene

The delocalised structure of benzene shows a benzene molecule as a hybrid state between Kekulé's two isomers with no separate single and double bonds.

In this hybrid state:

- each carbon atom contributes one electron from its p-orbital to form π-bonds
- the π-bonds are spread out or **delocalised** over the whole ring.

The delocalised model of a benzene molecule has identical carbon–carbon bonds making up the ring.

Key point from AS

- **Alkenes**
  Revise AS page 122

This model helps to explain the low reactivity of benzene compared with alkenes (see also p. 112).

Although the Kekulé structure is used for some purposes, the delocalised structure is a better representation of benzene. You will find both representations in books.

The double bond in an alkene is a **localised** π-bond between two carbon atoms. In the delocalised structure of benzene, each bond is identical with electron density spread out to encompass and stabilise the whole ring. The diagram below shows one of the π-bonds formed by delocalisation of electrons in a benzene molecule.

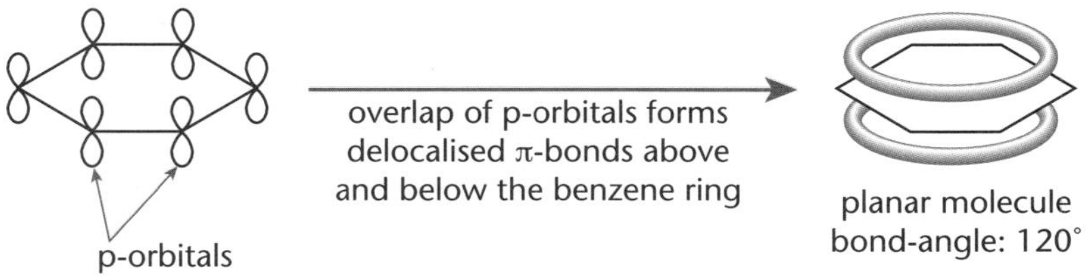

The delocalised structure of a benzene molecule is shown as a hexagon to represent the carbon skeleton and a circle to represent the six delocalised electrons.

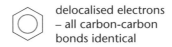

delocalised electrons
– all carbon-carbon
bonds identical

The shape of a benzene molecule

Key points from AS

- **Electron-pair repulsion theory**
  Revise AS pages 50–51

Using electron-pair repulsion theory:

- there are **three** centres of electron density surrounding each carbon atom
- the shape around each carbon atom is trigonal planar with bond angles of 120°.

This results in a benzene molecule that is **planar**.

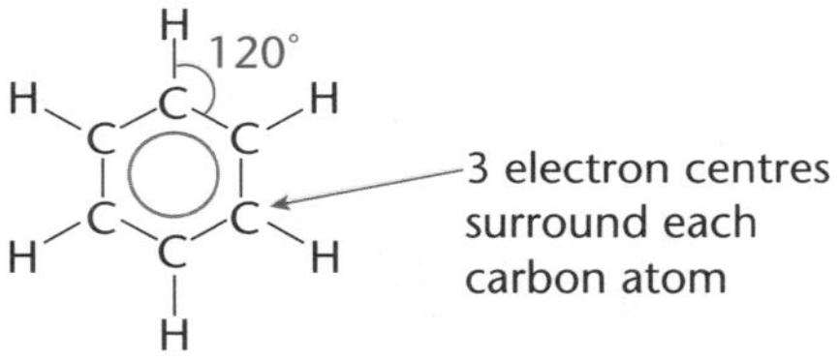

Experimental evidence for delocalisation

AQA	M4	NUFFIELD	M4
EDEXCEL	M5	SALTERS	M5
OCR	M4	WJEC	CH4
		NICCEA	M5

Bond length data

The Kekulé structure of benzene as cyclohexa-1,3,5-triene suggests two bond lengths for the separate single and double bonds.

- C—C bond length = 0.154 nm
- C=C bond length = 0.134 nm

Experiment shows only one C—C bond length of 0.139 nm, between the bond lengths for single and double carbon-carbon bonds.

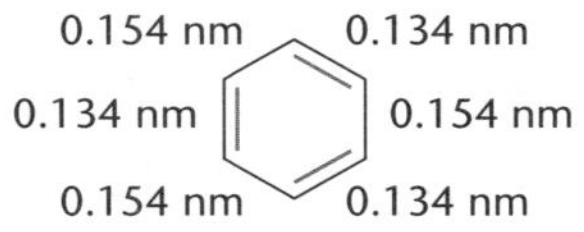

all C-C bonds are the same length: 0.139 nm

> This shows that each carbon-carbon bond in the benzene ring is intermediate between a single and a double bond.
>
> **KEY POINT**

Thermochemical evidence

Hydrogenation of cyclohexene

Each molecule of cyclohexene has **one** C=C double bond. The enthalpy change for the reaction of cyclohexene with hydrogen is shown below.

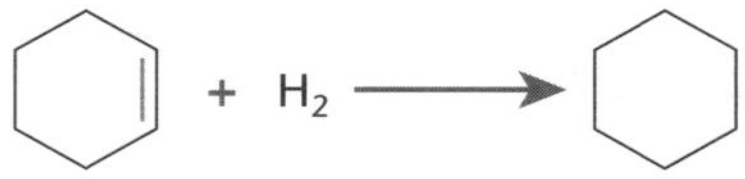

$\Delta H^{\ominus} = -120$ kJ mol^{-1}

Hydrogenation of benzene

The stability of benzene can also be demonstrated using thermochemical data for the reactions of bromine with cyclohexene and benzene.

The Kekulé structure of benzene as cyclohexa-1,3,5-triene has **three** double C=C bonds. It would be expected that the enthalpy change for the hydrogenation of this structure would be three times the enthalpy change for the **one** C=C bond in cyclohexene.

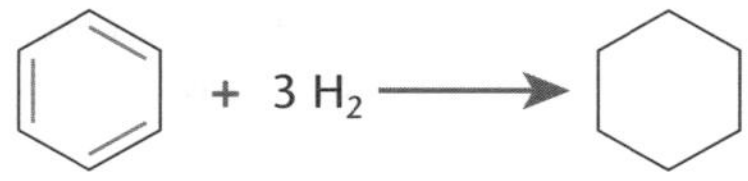

predicted enthalpy change:
$\Delta H^{\ominus} = 3 \times -120 = -360$ kJ mol^{-1}

- When benzene is reacted with hydrogen, the enthalpy change obtained is far less exothermic, $\Delta H^{\ominus} = -208$ kJ mol^{-1}.

KEY POINT

- The difference between the thermochemical data for cyclohexa-1,3,5-triene and benzene suggests that benzene has **more stable bonding** than the Kekulé structure.
- The delocalisation stability of benzene of –152 kJ mol^{-1} is the difference between the two enthalpy changes above. This is extra energy that must be provided to break the delocalised benzene ring.

Comparing electrophilic addition to arenes and alkenes

AQA	M4	SALTERS	M5
EDEXCEL	M5	WJEC	CH4
OCR	M4	NICCEA	M5
NUFFIELD	M4		

Key points from AS

- **Alkenes**
  Revise AS page 122
- **Addition reactions of alkenes**
  Revise AS pages 123–124

Arenes and alkenes both have π-bonds, but

- arenes have **delocalised** π-bonds
- alkenes have **localised** π-bonds
- the delocalised electron density in arenes is **less** than in alkenes.

Alkenes react with bromine in the dark at room temperature. Using the same conditions with benzene, there is no reaction. The stability of the delocalised system resists addition which would disrupt this stability.

Progress check

1 Name the **three** aromatic isomers of $C_6H_4Br_2$.

2 State two pieces of experimental evidence that support the delocalised structure of benzene.

3 Explain why alkenes, such as cyclohexene, are so much more reactive with electrophiles than arenes such as benzene.

1 1,2-dibromobenzene; 1,3-dibromobenzene; 1,4-dibromobenzene.

2 The enthalpy change of hydrogenation of benzene is less exothermic than three times the enthalpy change of hydrogenation of cyclohexene.
All the carbon-carbon bonds in the benzene ring are the same length.

3 Alkenes have localised π-bonds with a larger electron density than the delocalised π-bonds in benzene. The greater electron density is able to attract electrophiles more strongly. Also, the stability of the benzene ring must be disrupted if benzene is to react.

6.2 Reactions of arenes

After studying this section you should be able to:

- *describe electrophilic substitution of arenes: nitration, halogenation, alkylation, acylation and sulphonation*
- *describe the mechanism of electrophilic substitution in arenes*
- *understand the importance of reactions of arenes in the synthesis of commercially important materials*

LEARNING SUMMARY

Electrophilic substitution reactions of arenes

AQA	M4	SALTERS	M5
EDEXCEL	M5	WJEC	CH4
OCR	M4	NICCEA	M5
NUFFIELD	M4		

Many electrophiles react with alkenes by **addition**. However, electrophiles react with arenes by **substitution**, replacing a hydrogen atom on the ring. The difference in behaviour results from the high stability of the **delocalised** π-bonds in benzene compared with the **localised** π-bonds in alkenes (see pages 110–112).

- Benzene reacts with only very reactive electrophiles.
- The typical reaction of an arene is **electrophilic substitution.**

KEY POINT

In this section, reactions of benzene are discussed to illustrate electrophilic substitution reactions of arenes.

Nitration of arenes

AQA	M4	SALTERS	M5
EDEXCEL	M5	WJEC	CH4
OCR	M4	NICCEA	M5
NUFFIELD	M4		

The nitration of arenes produces aromatic nitro compounds, important for the synthesis of many important products including explosives and dyes (see page 121).

Nitration of benzene

Benzene is nitrated by concentrated nitric acid at 55°C in the presence of concentrated sulphuric acid, which acts as a catalyst.

Although some heat is required (55°C), too much may give further nitration of the benzene ring forming 1,3-dinitrobenzene.

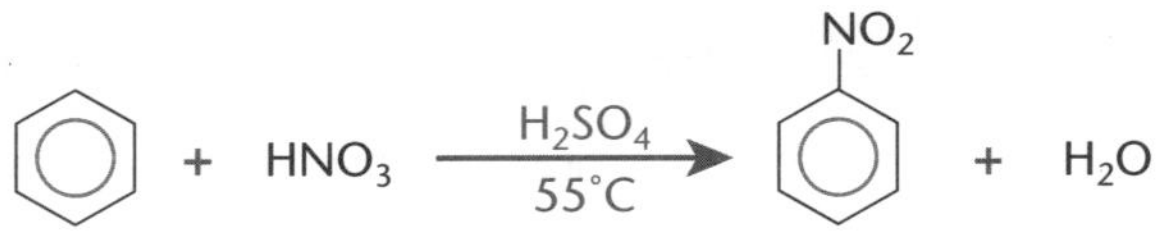

Mechanism

The nitronium ion is also called a **nitryl cation.**

- The role of the concentrated sulphuric acid is to generate the **nitronium ion**, NO_2^+, as the active electrophile:

$$HNO_3 + H_2SO_4 \longrightarrow H_2NO_3^+ + HSO_4^-$$
$$H_2NO_3^+ \longrightarrow NO_2^+ + H_2O$$

- The powerful NO_2^+ electrophile then reacts with benzene.

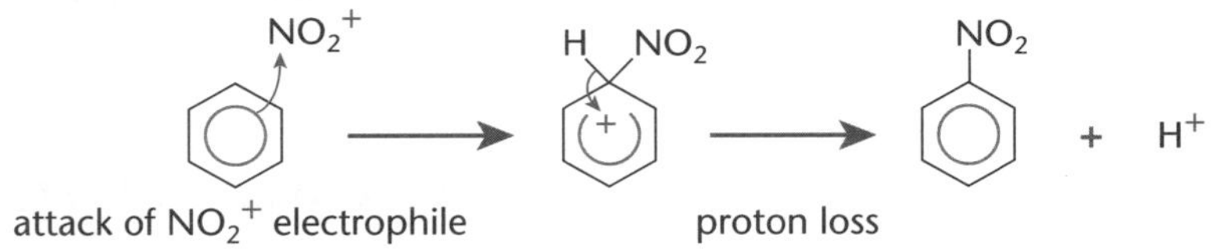

- The H^+ formed regenerates a molecule of H_2SO_4.

$$H^+ + HSO_4^- \longrightarrow H_2SO_4$$

The H_2SO_4 therefore acts as a **catalyst.**

- The H_2SO_4 molecule reacts with more nitric acid to form more nitronium ions.

Halogenation of arenes

EDEXCEL	M5	SALTERS	M5
OCR	M4	WJEC	CH4
NUFFIELD	M4	NICCEA	M5

Arenes react with halogens in the presence of a **halogen carrier**, which acts as a catalyst.

Suitable halogen carriers include:

- iron
- aluminium halides, e.g. $AlCl_3$ for chlorination.

Halogenation of benzene

The reaction takes place in warm conditions.

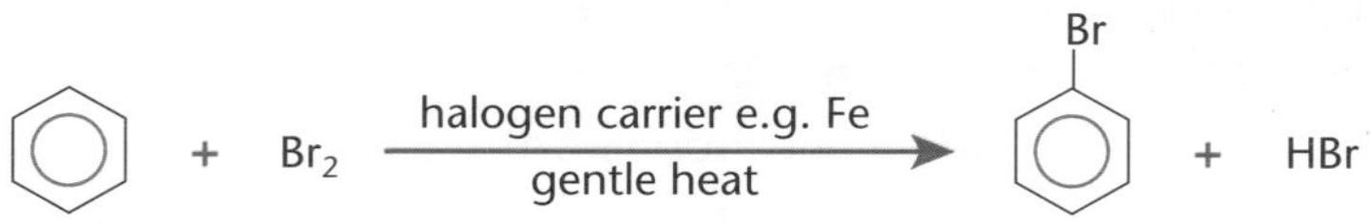

Mechanism (Edexcel only)

- Iron first reacts with bromine forming iron(III) bromide, $FeBr_3$.
- $FeBr_3$ acts as a halogen carrier polarising the Br—Br bond.

$$Br_2 + FeBr_3 \longrightarrow Br^{\delta+}{-}Br^{\delta-}\ FeBr_3$$

- This electrophile reacts with benzene.

This is a similar principle to the nitration of the benzene ring.

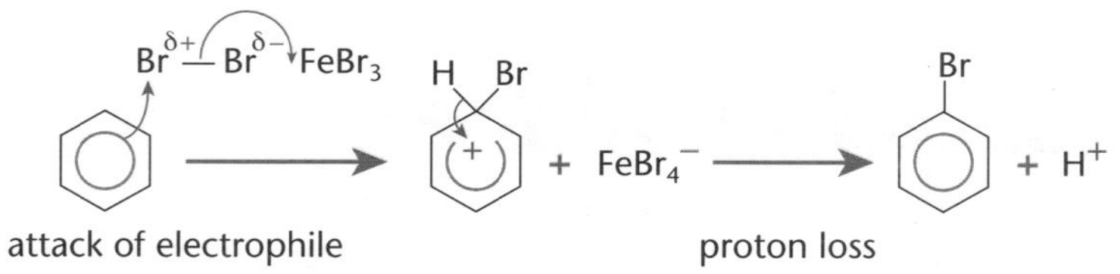

The $FeBr_3$ therefore acts as a **catalyst**.

- The H^+ formed generates $FeBr_3$.

$$H^+ + FeBr_4^- \longrightarrow FeBr_3 + HBr$$

- $FeBr_3$ can now polarise more bromine molecules.

Alkylation of arenes

AQA	M4	NUFFIELD	M4
EDEXCEL	M5	SALTERS	M5
OCR	M4		

Alkylation reactions of arenes are commonly known as *Friedel–Crafts* reactions. These are very important reactions industrially as they provide a means of introducing an alkyl group onto the benzene ring.

Alkylation of benzene

As with halogenation, a halogen carrier is required to generate a more reactive electrophile. With chloroethane, ethylbenzene is formed.

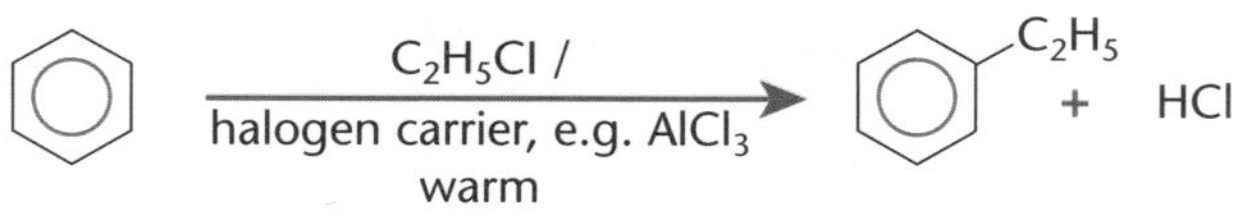

This is a similar principle to the halogenation of the benzene ring.

Mechanism (AQA and Edexcel only)

Using a tertiary halogenoalkane, RCl, in the presence of $AlCl_3$ as a halogen carrier, a **carbonium ion** is generated.

$$RCl + AlCl_3 \longrightarrow R^+ + AlCl_4^-$$

By using different halogenoalkanes, different alkyl groups can be substituted onto the benzene ring.

- This carbonium ion acts as a powerful electrophile which reacts with benzene.

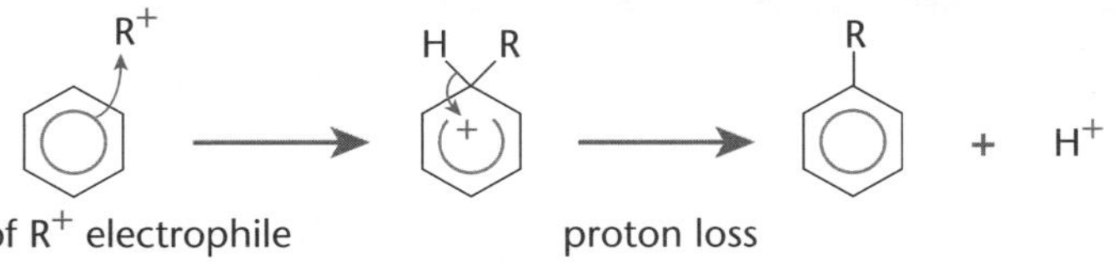

- The H^+ formed generates more $AlCl_3$.

$$H^+ + AlCl_4^- \longrightarrow AlCl_3 + HCl$$

- The $AlCl_3$ reacts with more of the tertiary halogenoalkane RCl to generate more carbonium ions. The **$AlCl_3$** acts as a **catalyst**.

Formation of ethylbenzene in industry

An important example of alkylation is the production of ethylbenzene, used to synthesise poly(phenylethene) (*polystyrene*). See page 126.

The reaction has been modified to make use of available starting materials.

Benzene and ethene are produced in large quantities in refineries.

Chloroethane is not readily available.

In industry, benzene is alkylated with ethene in the presence of HCl and $AlCl_3$ which both act as catalysts. Ethene is far cheaper and more readily available than chloroethane.

Chloroethane is first produced: $CH_2{=}CH_2 + HCl \longrightarrow CH_3CH_2Cl$

A complex forms: $CH_3CH_2Cl + AlCl_3 \longrightarrow CH_3CH_2^+\ AlCl_4^-$

Ethylbenzene forms: $C_6H_6 + CH_3CH_2^+AlCl_4^- \longrightarrow C_6H_5CH_2CH_3 + AlCl_3 + HCl.$

Note that $AlCl_3$ and HCl are regenerated and can react again.

Acylation of arenes

AQA M4 SALTERS M5
EDEXCEL M5

Acylation introduces an acyl group such as $CH_3C{=}O$ onto the benzene ring. An acyl chloride is used in the presence of a halogen carrier.

The mechanism for acylation is similar to that of alkylation.

The acylium ion is then able to react with the benzene ring, introducing an acyl group onto the ring.

Acylation of benzene

The acylation of benzene is another example of a *Friedel–Crafts* reaction and is also important industrially.

Using ethanoyl chloride, CH_3COCl, the ethanoyl group $CH_3C{=}O$ can be introduced onto the benzene ring.

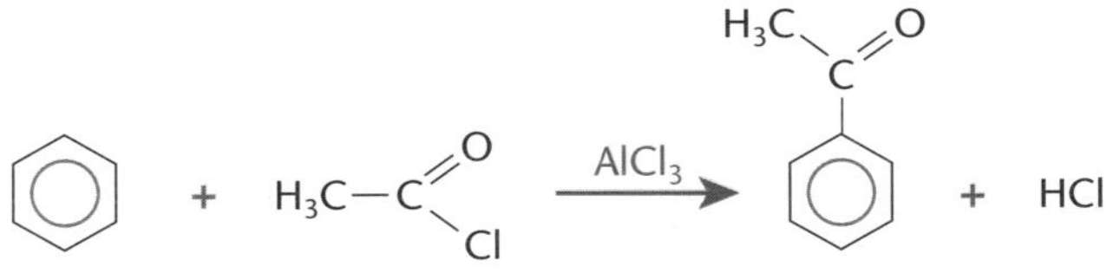

This reaction takes place by a similar mechanism to the alkylation reaction above (Edexcel only).

Sulphonation of arenes

NUFFIELD M4 SALTERS M5

Sulphonation of arenes produces sulphonic acids. Sodium salts of sulphonic acids are used for detergents and fabric conditioners.

Sulphonation of benzene

Benzene reacts with fuming sulphuric acid (concentrated sulphuric acid saturated with sulphur trioxide) forming a sulphonic acid.

The reaction takes place between benzene and sulphur trioxide.

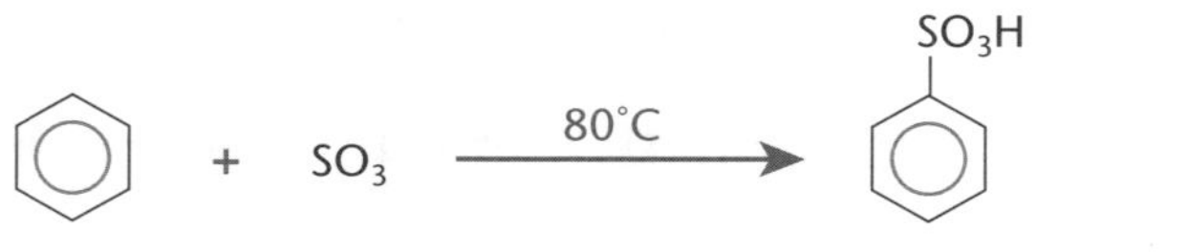

benzenesulphonic acid

Oxidation of substituted arenes

EDEXCEL M5 WJEC CH4

The side chains of substituted arenes can be readily oxidised by prolonged reflux with a strong oxidising agent.

Oxidation of methylbenzene

Using alkaline potassium manganate(VII), followed by addition of dilute aqueous acid, methylbenzene is oxidised to benzenecarboxylic acid **(benzoic acid)**.

benzenecarboxylic acid

Progress check

1 What is the common type of reaction of arenes?

2 Benzene reacts with bromine, nitric acid and chloroethane.
(a) Using C_6H_6 to represent benzene and C_6H_5 to represent the phenyl group, write balanced equations for each of these reactions.
(b) State the essential conditions that are needed for each of these reactions.

1 Electrophilic substitution.

2 (a) $C_6H_6 + Br_2 \longrightarrow C_6H_5Br + HBr$
$C_6H_6 + HNO_3 \longrightarrow C_6H_5NO_2 + H_2O$
$C_6H_6 + C_2H_5Br \longrightarrow C_6H_5C_2H_5 + HBr$
(b) With bromine, a halogen carrier such as Fe.
With nitric acid, concentrated sulphuric acid at 55°C.
With chloroethane, a halogen carrier such as Fe.

6.3 Phenols

After studying this section you should be able to:

- *state the uses of phenols in antiseptics and disinfectants*
- *describe the reactions of phenol with sodium and bases*
- *explain the ease of bromination of phenol compared with benzene*

LEARNING SUMMARY

Phenols

General structure: ArOH (Ar is an aromatic ring)

EDEXCEL M5 | NUFFIELD M4
OCR M4 | WJEC CH4

Phenols have an aromatic ring bonded directly to a hydroxyl (–OH) group. Some phenols are shown below.

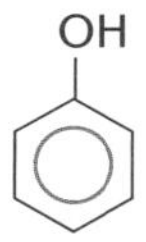

phenol, C_6H_5OH

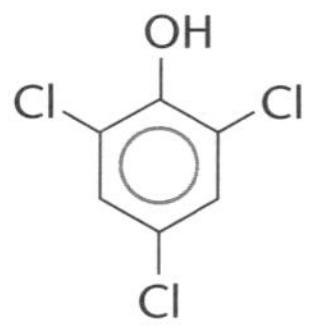

2,4,6-trichlorophenol (TCP)

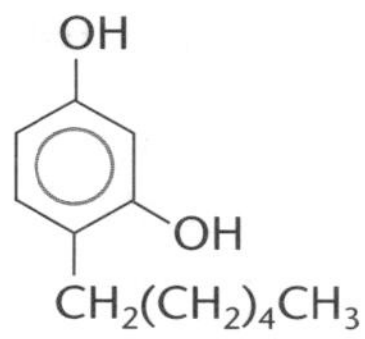

4-hexyl-3-hydroxyphenol (in cough drops)

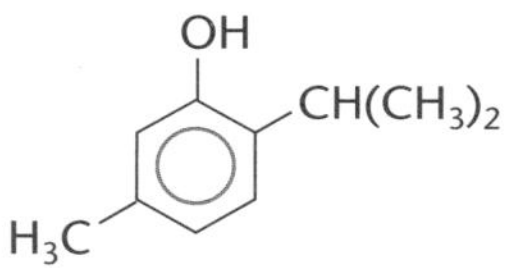

thymol (in thyme)

Uses

Dilute solutions of phenol are used as disinfectants and phenol was probably the first antiseptic. However, phenol is toxic and can cause burns to the skin. Less toxic phenols are used nowadays in antiseptics.

Phenols are also used in the production of plastics, such as the widely used phenol-formaldehyde resins. More complex phenols, such as thymol shown above, can be used as flavourings and aromas and these are obtained from essential oils of plants.

Properties

Like alcohols, a phenol has a **hydroxyl group** which is able to form intermolecular **hydrogen bonds**. The resulting properties include:

- higher melting and boiling points than hydrocarbons with similar relative molecular masses
- some solubility in water.

Reactions of phenol as an acid

EDEXCEL M5 | NUFFIELD M4
OCR M4 | WJEC CH4

Unlike the neutrality shown by alcohols, phenols are very weak acids. The hydroxyl group donates a proton, H^+, by breaking the O–H bond.

$$C_6H_5OH \rightleftharpoons H^+ + C_6H_5O^-$$

Phenols takes part in some typical acid reactions, reacting with sodium and bases to form ionic salts called **phenoxides**. However, the acidity is too weak to release carbon dioxide from carbonates (see carboxylic acids, page 100).

Key points from AS

- **Tests for alcohols**
  Revise AS page 131

Reaction with sodium

As with the hydroxyl groups in water and ethanol, phenol reacts with sodium, releasing hydrogen gas and forming a solution of sodium phenoxide, $C_6H_5O^-Na^+$.

$$2C_6H_5OH + 2Na \longrightarrow 2C_6H_5ONa + H_2$$

Compare these reactions with those of carboxylic acids p. 100.

Reaction with an alkali

Phenol is a weak acid and is neutralised by aqueous alkalis. For example, aqueous sodium hydroxide forms the salt sodium phenoxide.

$$C_6H_5OH + NaOH \longrightarrow C_6H_5ONa + H_2O$$

Electrophilic substitution of the aryl ring of phenol

EDEXCEL M5 NUFFIELD M4
OCR M4 WJEC CH4

The aryl ring of phenol has a greater electron density than benzene because the adjacent oxygen reinforces the electron density in the ring.

This results in:

- a greater electron density in the aryl ring
- greater reactivity of the ring towards electrophiles.

Bromination of phenol

- Phenol reacts directly with bromine. However Benzene reacts with bromine only in the presence of a halogen carrier (see page 114).
- Phenol undergoes multiple substitution with bromine. Benzene is monosubstituted only.

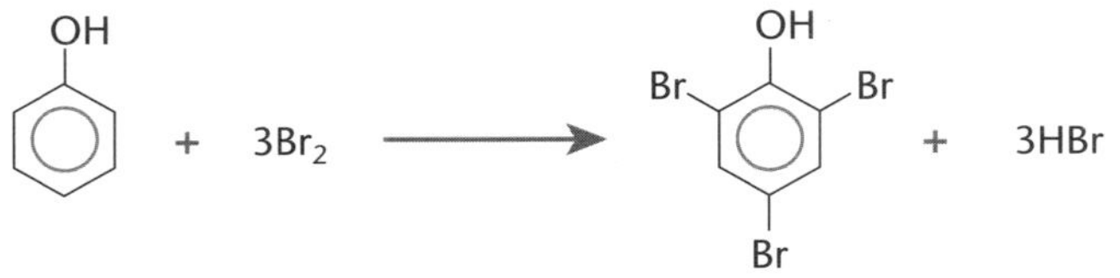

- The organic product, 2,4,6-tribromophenol separates as a white solid.
- The bromine is decolourised.

Formation of esters

EDEXCEL M5 WJEC CH4

Phenols form esters with carboxylic acids **only** in the presence of aqueous alkali.

The negative inductive effect from the aryl ring reduces the electron density of the phenolic –OH group and its effectiveness as a nucleophile. An alkali increases the electron density by generating the phenoxide ion:

$$OH^- + C_6H_5OH \longrightarrow C_6H_5O^- + H_2O$$
$$CH_3COOH + C_6H_5O^- \longrightarrow CH_3COOC_6H_5 + OH^-$$

Phenols readily form esters with acyl chlorides.

E.g. $CH_3COCl + C_6H_5OH \longrightarrow CH_3COOC_6H_5 + HCl$

Progress check

Phenols are easily nitrated with dilute nitric acid in the cold. As with bromine, multiple substitution takes place.
Suggest an equation for the reaction that takes place.

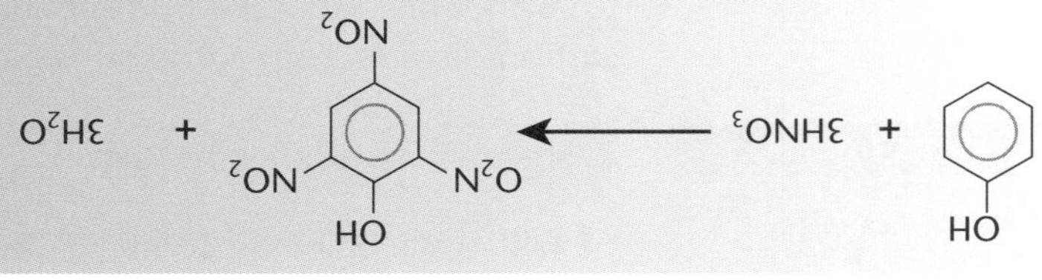

6.4 Amines

After studying this section you should be able to:

- *explain the relative basicities of ethylamine and phenylamine*
- *describe the reactions of primary amines with acids to form salts*
- *describe the formation of phenylamine by reduction of nitrobenzene*
- *describe the synthesis of an azo-dye from phenylamine*
- *describe the reactions of amines with halogenoalkanes and acyl chlorides*

LEARNING SUMMARY

Aliphatic and aromatic amines

AQA	M4	SALTERS	M4
EDEXCEL	M4	WJEC	CH4
OCR	M4	NICCEA	M5
NUFFIELD	M6		

Amines are organic compounds containing nitrogen derived from ammonia, NH_3. Amines are classified as primary, secondary or tertiary amines depending on how many of the hydrogen atoms in ammonia have been replaced by organic groups. The diagram below shows examples of aliphatic and aromatic amines.

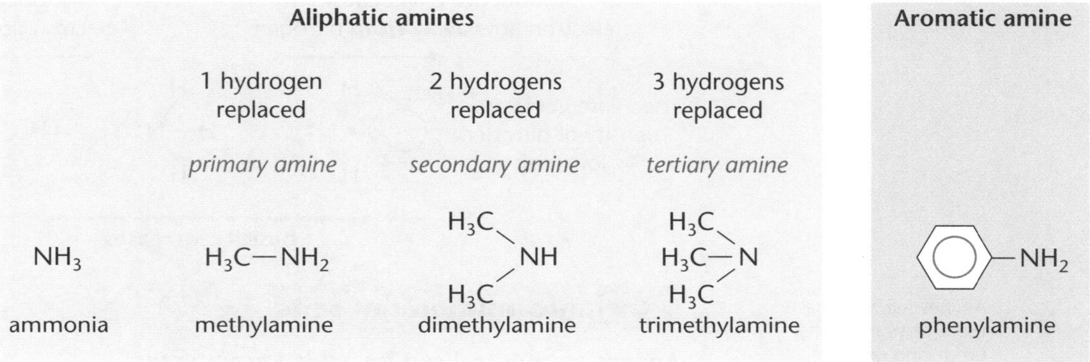

Amines in nature

Amines are found commonly in nature. They are weak bases and the shortest chain amines, e.g. ethylamine $C_2H_5NH_2$, smell of fish. Some diamines, such as putrescine $H_2N(CH_2)_4NH_2$, and cadaverine $H_2N(CH_2)_5NH_2$, are found in decaying flesh.

Polarity

Notice the similarity with ammonia.

The presence of an electronegative nitrogen atom in amines results in polar molecules. The diagram shows the polarity of the primary amine methylamine.

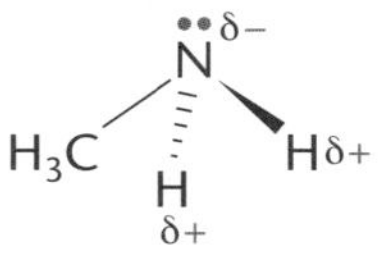

Physical properties of amines

As with other polar functional groups, the solubility of amines in water decreases with increasing carbon chain length as the non-polar contribution to the molecule becomes more important.

The amino group dominates the physical properties of short-chain amines.

Hydrogen bonding takes place between amine molecules, resulting in:

- higher melting and boiling points than alkanes of comparable relative molecular mass
- solubility in water – amines with fewer than six carbons mix with water in all proportions.

Amines as bases

AQA	M4	SALTERS	M4
EDEXCEL	M4	WJEC	CH4
OCR	M4	NICCEA	M5
NUFFIELD	M6		

Amines are the organic bases.

Amines are weak bases because they only partially associate with protons:

$$RNH_2 + H^+ \rightleftharpoons RNH_3^+$$

The basic strength of an amine measures its ability to **accept** a proton, H^+. This depends upon:

- the size of the δ– charge on the amino nitrogen atom
- the availability of the nitrogen lone pair.

The basicity of aliphatic and aromatic amines

Aliphatic amines are stronger bases than ammonia; aromatic amines are substantially weaker.

KEY POINT

- Electron-donating groups, e.g. alkyl groups, **increase** the basic strength.
- Electron-withdrawing groups, e.g. C_6H_5, **decrease** the basic strength.

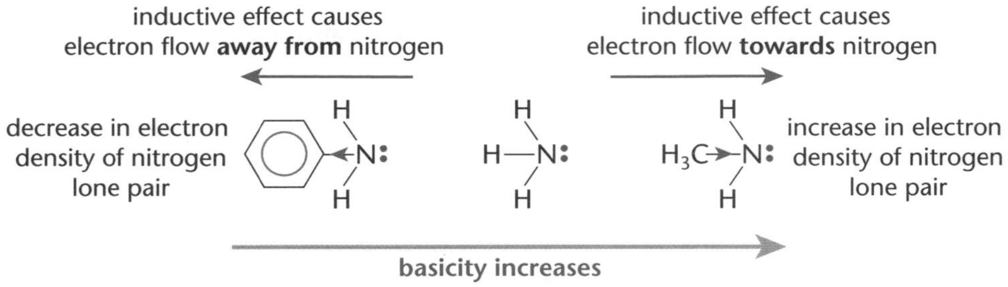

Amines react by the usual 'base reactions' producing organic ammonium salts.

A proton, H^+, is added to the amino nitrogen atom.

Organic ammonium salts

Amines are **neutralised by acids** forming salts.

In the example below, methylamine is neutralised by hydrochloric acid forming a primary **ammonium salt**.

$$CH_3NH_2 + HCl \longrightarrow CH_3NH_3^+Cl^-$$

methylammonium chloride

On evaporation of water, the primary ammonium salt crystallises out as an ionic compound.

Preparation of amines

AQA	M4	WJEC	CH4
EDEXCEL	M4, M5	NICCEA	M5
OCR	M4		

Other reducing agents can also be used:

- H_2/Ni catalyst
- $LiAlH_4$ in dry ether

$LiAlH_4$ is an almost universal reducing agent and works in most examples of organic reduction.

Preparation of aliphatic amines from nitriles

Aliphatic amines can be prepared by **reducing** a nitrile with a suitable reducing agent (e.g. Na in ethanol; $LiAlH_4$ in dry ether). (See also page 96.)

$$CH_3CH_2C{\equiv}N + 4[H] \longrightarrow CH_3CH_2CH_2NH_2$$

Preparation of aromatic amines from nitroarenes

Aromatic amines can be prepared by **reducing** a nitroarene. For example nitrobenzene, is reduced by Sn in concentrated HCl to form phenylamine.

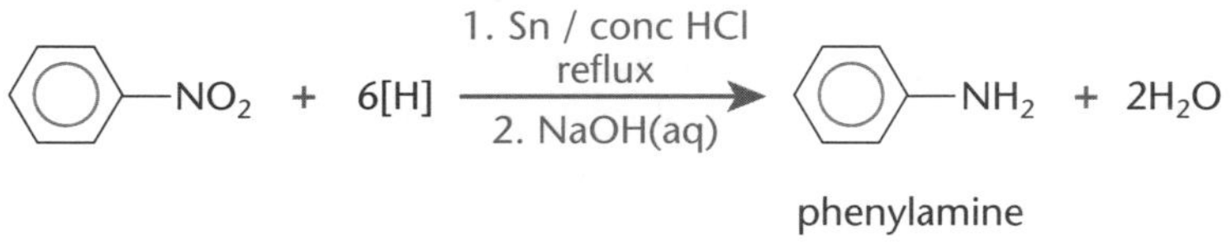

- Using tin and hydrochloric acid, the salt $C_6H_5NH_3^+Cl^-$ is formed.
- The amine is obtained from this salt by adding aqueous alkali.

This process is often called the *Hofmann degradation* and can be used to move down a homologous series.

See also: Increasing the carbon chain length p. 95.

Preparation of amines from amides (Edexcel only)

Primary amines can be prepared by heating a primary amide with bromine in alkali. The amine formed has **one less carbon** atom than the starting amide.

$$RCONH_2 + Br_2 + 4NaOH \longrightarrow RNH_2 + Na_2CO_3 + 2NaBr + 2H_2O$$

The preparation of dyes from aromatic amines

EDEXCEL	M5	WJEC	CH4
OCR	M4	NICCEA	M5
SALTERS	M5		

Aromatic amines, such as phenylamine, are important industrially for the production of dyes. Modern dyes are formed in a two-stage synthesis:

- the aromatic amine is converted into a **diazonium salt**
- the diazonium salt is **coupled** with an aromatic compound such as phenol, forming an **azo dye**.

Formation of diazonium salts

An aromatic amine, such as phenylamine, forms a diazonium salt in the presence of nitrous acid, HNO_2, and hydrochloric acid.

- HNO_2 is unstable and is prepared *in situ* from $NaNO_2$ and HCl(aq).

$$NaNO_2 + HCl \longrightarrow HNO_2 + NaCl$$

- This mixture is then reacted with phenylamine. It is important to keep the temperature **below 10°C** because diazonium salts decompose above this temperature.

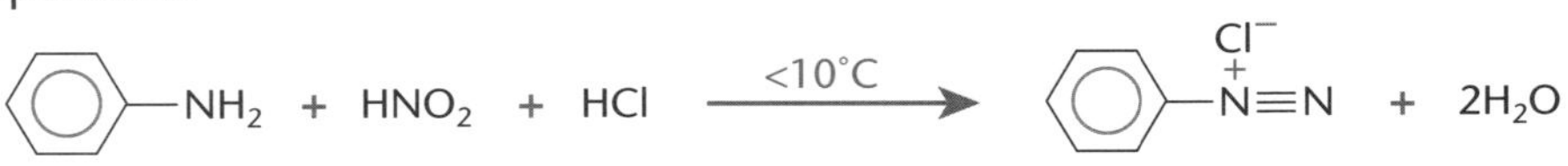

benzenediazonium chloride
(diazonium salt)

$$C_6H_5NH_2 + HNO_2 + HCl \longrightarrow C_6H_5N_2^+Cl^- + 2H_2O$$

Formation of azo dyes by coupling

The diazonium salt is coupled with a suitable aromatic compound **below 10°C** in **aqueous alkali** to produce an azo dye.

E.g. coupling of benzenediazonium chloride with phenol:

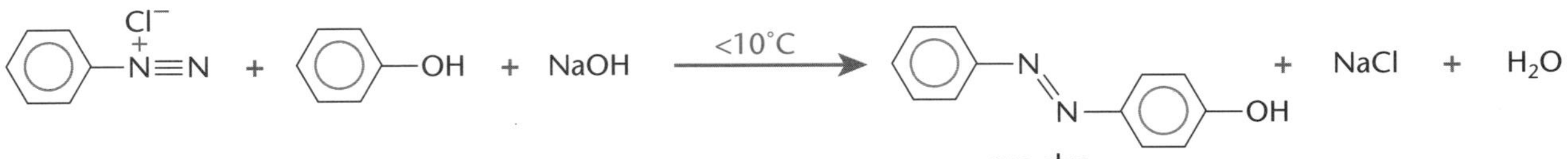

azo dye

- The azo dye is coloured.
- Coupling of a diazonium salt with different aromatic compounds forms dyes with different colours.

Using benzene for the synthesis of dyes

Benzene is used as the raw material for the synthesis of dyes.
A four-stage synthesis is shown below:

- benzene is nitrated to nitrobenzene (see page 113)
- nitrobenzene is reduced to phenylamine (see above)
- phenylamine is converted into a **diazonium salt**
- the diazonium salt is **coupled** with an aromatic compound such as phenol, forming an **azo dye**.

Reactions of amines with halogenoalkanes

AQA	M4	WJEC	CH4
NUFFIELD	M6	NICCEA	M5

Key points from AS

- **Nucleophilic substitution reactions of halogenoalkanes**
  Revise AS pages 132–133

Formation of a primary amine by nucleophilic substitution

Ammonia and amines react as **nucleophiles** with halogenoalkanes in substitution reactions.

- **Excess ammonia** reacts with halogenoalkanes in hot ethanol, forming a primary ammonium salt.

$$RX + NH_3 \longrightarrow RNH_3^+X^-$$

- Proton transfer with ammonia forms a **primary amine** RNH_2

$$RNH_3^+X^- + NH_3 \longrightarrow RNH_2 + NH_4X$$

Further substitution (AQA only)

At each stage, a hydrogen atom is replaced by an alkyl group.

- With an **excess** of the **halogenoalkane**, further substitution takes place which continues until a quaternary ammonium salt is obtained.

$$RNH_2 \xrightarrow{RX} \underset{\text{secondary ammonium salt}}{R_2NH_2^+X^-} \xrightarrow{RX} \underset{\text{tertiary ammonium salt}}{R_3NH^+X^-} \xrightarrow{RX} \underset{\text{quaternary ammonium salt}}{R_4N^+X^-}$$

Each reaction stage proceeds by **nucleophilic substitution**. The mechanism for the final stage in the formation of the quaternary salt tetramethylammonium bromide is shown below.

Quaternary ammonium salts are used as cationic surfactants in fabric softeners.

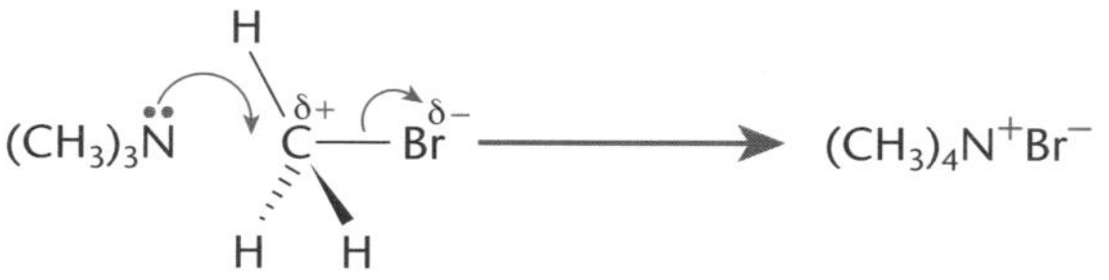

Reactions of amines with acyl chlorides

AQA	M4	SALTERS	M4
EDEXCEL	M4	WJEC	CH4
NUFFIELD	M6	NICCEA	M5

Amines react with acyl chorides forming amides.

E.g.: $RCOCl + CH_3NH_2 \longrightarrow RCONHCH_3 + HCl$

$CH_3NH_2 + HCl \longrightarrow CH_3NH_3^+Cl^-$

For details, see also acyl chlorides p.104.

Progress check

1 Explain why ethylamine is a stronger base than phenylamine.

2 (a) Write equations for the conversion of:
   (i) benzene to nitrobenzene
   (ii) nitrobenzene to phenylamine
   (iii) phenylamine to benzenediazonium chloride.

3 (a) Write down the formula of the organic product formed between:
   (i) bromoethane and excess ammonia
   (ii) ammonia and excess bromoethane.

1 The ethyl group has a positive inductive effect which reinforces the electron density of the nitrogen atom of the amine. This lone pair on the nitrogen will attract protons more strongly.
The phenyl group has a negative inductive effect which reduces the electron density on the nitrogen atom of the amine. This lone pair on the nitrogen will attract protons less strongly.

2 (a) (i) $C_6H_6 + HNO_3 \longrightarrow C_6H_5NO_2 + H_2O$
(ii) $C_6H_5NO_2 + 6[H] \longrightarrow C_6H_5NH_2 + 2H_2O$
(iii) $C_6H_5NH_2 + HNO_2 + HCl \longrightarrow C_6H_5N_2^+Cl^- + 2H_2O$

3 (a) (i) $C_2H_5NH_2$ (ii) $(C_2H_5)_4N^+Br^-$

6.5 Amino acids

After studying this section you should be able to:

- describe the acid–base properties of amino acids and the formation of zwitterions
- explain the formation of polypeptides and proteins as condensation polymers of amino acids
- describe the acid hydrolysis of proteins and peptides

LEARNING SUMMARY

Amino acids

AQA	M4	SALTERS	M4
EDEXCEL	M4	WJEC	CH4
OCR	M4	NICCEA	M5
NUFFIELD	M6		

There are 22 naturally occurring amino acids.

A typical amino acid has organic molecules with both acidic and basic properties comprising:

- a basic amino group, $-NH_2$
- an acidic carboxyl group, $-COOH$.

Each amino acid has a unique organic R group or side chain. The formula of a typical amino acid is shown below.

$$H_2N-\underset{\underset{H}{|}}{\overset{\overset{R}{|}}{C}}-COOH$$

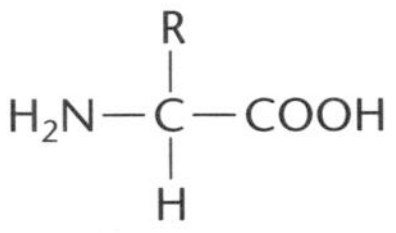

Examples of some amino acids are shown below:

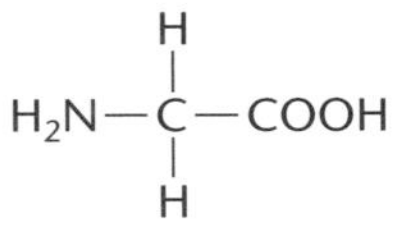

$$H_2N-\underset{\underset{H}{|}}{\overset{\overset{H}{|}}{C}}-COOH \qquad H_2N-\underset{\underset{H}{|}}{\overset{\overset{CH_3}{|}}{C}}-COOH \qquad H_2N-\underset{\underset{H}{|}}{\overset{\overset{CH(CH_3)_2}{|}}{C}}-COOH$$

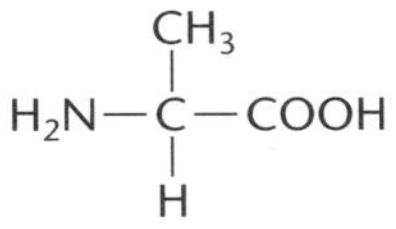

R = H   R = CH_3   R = CH(CH_3)_2
glycine   *alanine*   *valine*

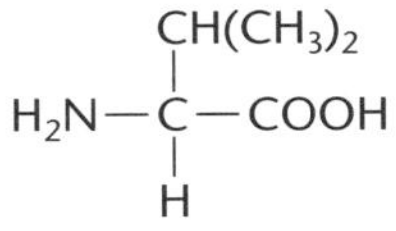

Acid–base properties of amino acids

AQA	M4	SALTERS	M4
EDEXCEL	M4	WJEC	CH4
OCR	M4	NICCEA	M5
NUFFIELD	M6		

Isoelectric points and zwitterions

Each amino acid has a particular pH called the **isoelectric point** at which the overall charge on an amino acid molecule is zero.

Examples of isoelectric points

amino acid	aspartic acid	glycine	histidine	arginine
isoelectric point	3.0	6.1	7.6	10.8

At the isoelectric point, the amino acids exist in equilibrium with its zwitterion form.

At the isoelectric point, an amino acid exists as a zwitterion:

- the **carboxyl** group **has donated** a proton to the **amino** group which form a positive NH_3^+ ion.

A **zwitterion** is a dipolar ion with both positive and negative charges in different parts of the molecule.

$$H_3\overset{+}{N}-\underset{\underset{H}{|}}{\overset{\overset{R}{|}}{C}}-C\overset{\displaystyle O}{\underset{\displaystyle O^-}{<}}$$

zwitterion – two ions
in one molecule

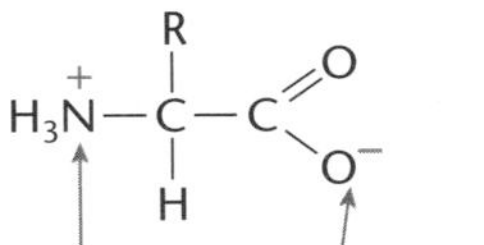

Amino acids as bases

In strongly **acidic** conditions a **positive ion** forms:

- an amino acid behaves as a **base**
- the COO^- ion gains a proton.

> Because of their reactions with strong acids and strong bases, amino acids act as buffers and help to stabilise the pH of living systems.

$$H_3\overset{+}{N}-\underset{\underset{H}{|}}{\overset{\overset{R}{|}}{C}}-C\overset{\diagup O}{\diagdown O^-} + H^+ \longrightarrow H_3\overset{+}{N}-\underset{\underset{H}{|}}{\overset{\overset{R}{|}}{C}}-C\overset{\diagup O}{\diagdown OH}$$

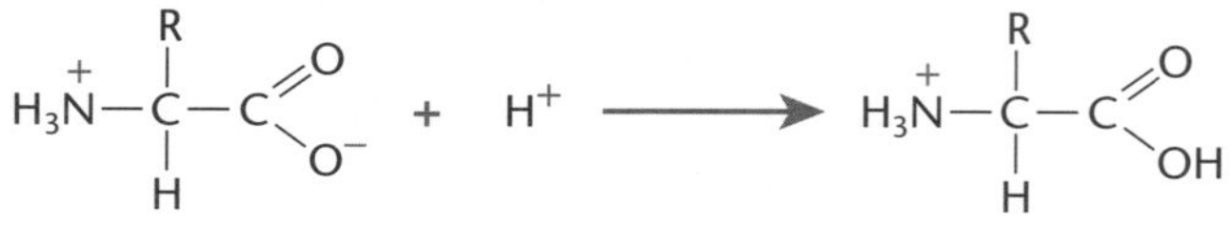

positive ion

Amino acids as acids

In strongly **alkaline** conditions a **negative ion** forms:

- an amino acid behaves as an **acid**
- the NH_3^+ ion loses a proton.

$$H_3\overset{+}{N}-\underset{\underset{H}{|}}{\overset{\overset{R}{|}}{C}}-C\overset{\diagup O}{\diagdown O^-} + OH^- \longrightarrow H_2N-\underset{\underset{H}{|}}{\overset{\overset{R}{|}}{C}}-C\overset{\diagup O}{\diagdown O^-} + H_2O$$

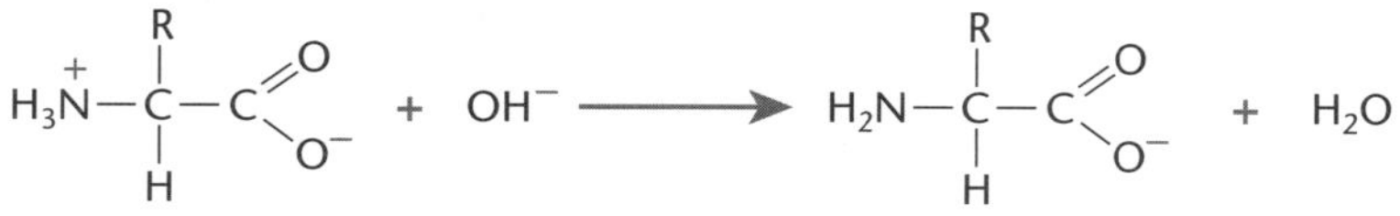

negative ion

> **KEY POINT**
> - At the **isoelectric point**, the amino acid is **neutral**.
> - At a pH more **acidic** than the isoelectric point, the amino acid forms a **positive ion**.
> - At a pH more **alkaline** than the isoelectric point, the amino acid forms a **negative ion**.

Physical properties of amino acids

AQA	M4	SALTERS	M4
EDEXCEL	M4	WJEC	CH4
OCR	M4	NICCEA	M5
NUFFIELD	M6		

Solid amino acids

Solid amino acids have higher melting points than expected from their molecular masses and structure. This suggests that amino acids crystallise in a giant lattice with strong electrostatic forces between the **zwitterions**.

Optical isomers

With the exception of aminoethanoic acid (*glycine*), H_2NCH_2COOH, all amino acids have a chiral centre and are optically active (see page 92).

Polypeptides and proteins

AQA	M4	SALTERS	M4
OCR	M4	WJEC	CH4
NUFFIELD	M6	NICCEA	M5

In nature, individual amino acids are linked together in chains as **polypeptides** and **proteins** (see page 127).

The diagram below shows the condensation of the amino acids glycine and alanine to form a **dipeptide**. The amino acids are bonded together by a **peptide link**.

> A protein is formed by condensation polymerisation of amino acids. See p. 127.

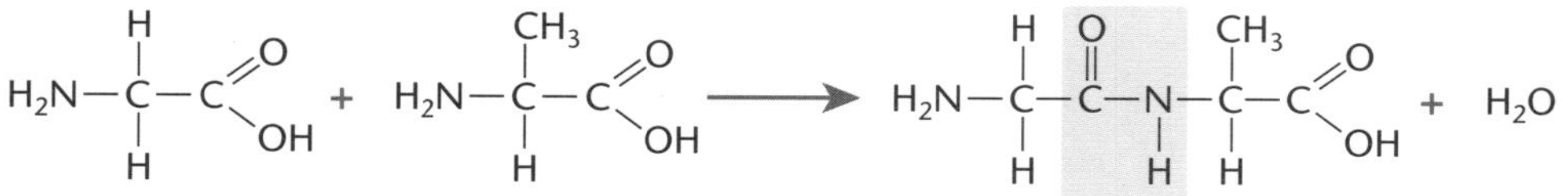

glycine   alanine   peptide link

A polypeptide is the name given to a short chain of amino acids linked by peptide bonds.

A protein is simply the name given to a long-chain polypeptide.

Each peptide link forms:

- between the **carboxyl group** of glycine and the **amino group** of alanine
- with loss of a water molecule in a **condensation reaction**.

Further condensation reactions between amino acids build up a **polypeptide** or **protein**.

- For each amino acid added to a protein chain one water molecule is lost.
- Most common proteins contain more than 100 amino acids.
- Each protein has a unique sequence of amino acids and a complex three-dimensional shape held together by intermolecular bonds including hydrogen bonds.

Hydrolysis of proteins

AQA	M4	NUFFIELD	M6
OCR	M4	SALTERS	M4

Hydrolysis breaks down a protein into its separate amino acids.

- The protein is refluxed with 6 mol dm^{-3} HCl(aq) for 24 hours.
- The resulting solution is neutralised.

The equation below shows the hydrolysis of a dipeptide.

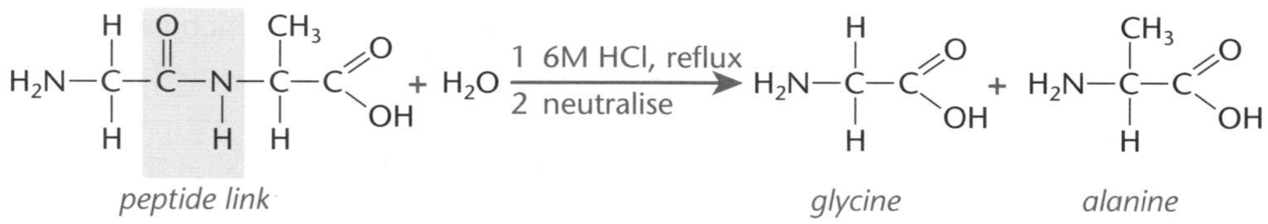

- Hydrolysis of a protein is the reverse process to the condensation polymerisation that forms proteins.
- Biological systems use enzymes to catalyse this hydrolysis, which takes place at body temperature without the need for acid or alkali.

Identifying the amino acids in a protein

To determine the amino acids present in a protein, the protein is first boiled with 6 mol dm^{-3} hydrochloric acid. The amino acids formed are separated using paper chromatography and made visible by spraying the paper with ninhydrin.

Each amino acid moves a different distance on the chromatography paper, making for easy identification of the amino acids in the protein.

Progress check

1 The isoelectric point of serine (R = $-CH_2OH$) is 5.7. Draw the form of the molecule in aqueous solutions of pH 3.0, pH 5.7 and pH 10.0.

2 Draw the structure of the tripeptide with the sequence alanine–serine–aspartic acid. (alanine: R = $-CH_3$; aspartic acid: R = $-CH_2COOH$).

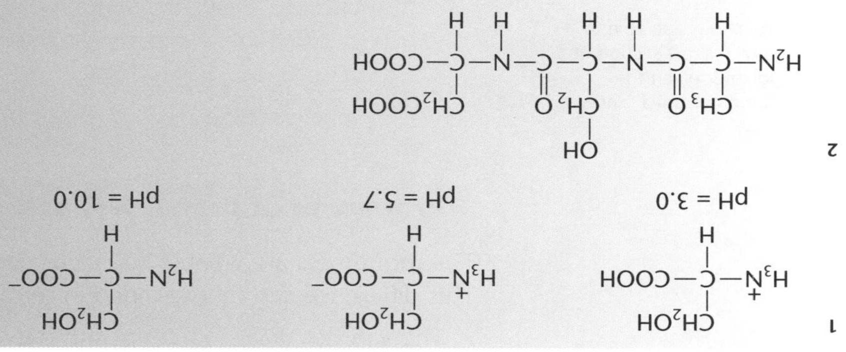

6.6 Polymers

After studying this section you should be able to:

- *describe the characteristics of addition polymerisation*
- *describe the characteristics of condensation polymerisation in polypeptides and proteins, polyamides and polyesters*
- *discuss the disposal of polymers*

LEARNING SUMMARY

During the study of alkenes in AS Chemistry, you learnt about addition polymers. For A2 Chemistry, addition polymerisation is reviewed and compared with condensation polymerisation.

Monomers and polymers

AQA	M4	SALTERS	M4
EDEXCEL	M5	WJEC	CH4
OCR	M4	NICCEA	M4
NUFFIELD	M6		

A **polymer** is a compound comprising very large molecules that are multiples of simpler chemical units called **monomers**.

Monomers are small molecules which can combine together to form a single large molecule, called a **polymer**.

Two processes lead to formation of a polymer.

- **Addition polymerisation** – monomers react together forming the polymer **only**. There are no by-products.
- **Condensation polymerisation** – monomers react together forming the polymer **and** a simple compound, usually water.

KEY POINT

Addition polymerisation

AQA	M4	NUFFIELD	M6
EDEXCEL	M5	SALTERS	M4
OCR	M4	WJEC	CH4

Key points from AS

- **Addition polymerisation of alkenes**
  Revise AS page 125

Notice that **n** monomer molecules produce **one** polymer molecule with **n** repeat units.

You should be able to draw a short section of a polymer given the monomer units (and vice versa).

In addition polymerisation:

- the monomer is an **unsaturated** molecule with a double **C=C** bond
- the double bond is **lost** as the **saturated** polymer forms.

Many different addition polymers can be formed using different monomer units based upon alkenes.

monomer unit **repeat unit of polymer**

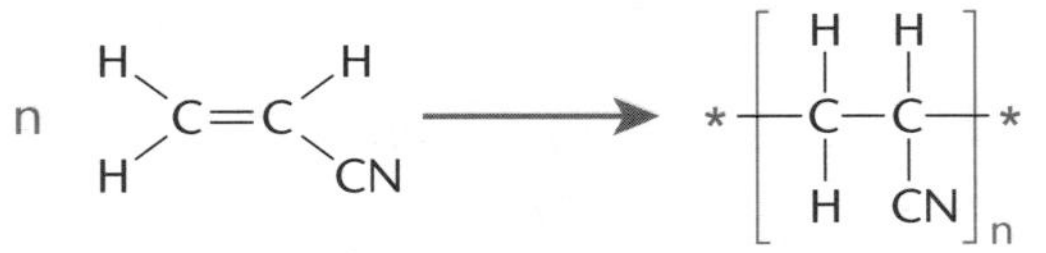

propenenitrile poly(propenenitrile) *Orlon*

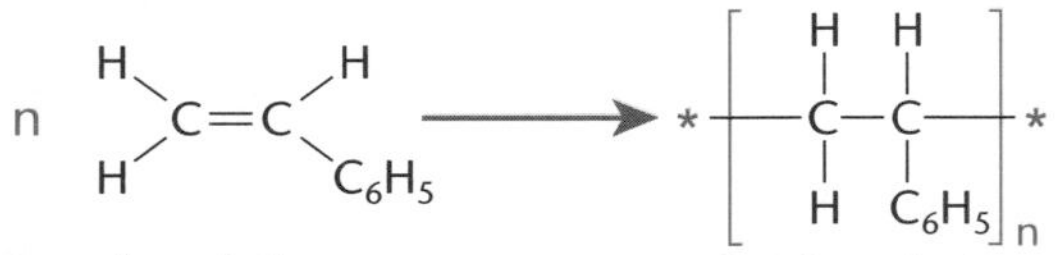

phenylethene poly(phenylethene) *polystyrene*

Properties of monomers and polymers

The monomers are volatile liquids or gases. Polymers are solids.
This difference can be explained in terms of van der Waals' forces.

- The van der Waals' forces acting between the large polymer molecules are much stronger than those acting between the much smaller monomer molecules.

Condensation polymerisation

AQA	M4	SALTERS	M4
EDEXCEL	M4, M5	WJEC	CH4
OCR	M4	NICCEA	M4, M5
NUFFIELD	M6		

In condensation polymerisation, the formation of a bond between monomer units also produces a small molecule such as H_2O or HCl.

Condensation polymers can be divided into natural polymers and man-made (synthetic) polymers.

- Natural polymers in living organisms include proteins, cellulose, rayon and DNA.
- Man-made polymers include synthetic fibres such as polyamides (e.g. *nylon*) and polyesters (e.g. *terylene*).

Proteins and polypeptides

A protein is a chain of many amino acids linked together with 'peptide bonds' formed by **condensation polymerisation**.

Each amino acid molecule has both **amino** and **carboxyl** groups.
A **peptide** bond forms, with loss of a water molecule, between:

- the **amino** group of one amino acid molecule and
- the **carboxyl** group of another amino acid molecule.

The diagram below shows how amino acid molecules are linked together during condensation polymerisation.

The formation of proteins from amino acids is also discussed in more detail on pp. 124–125.

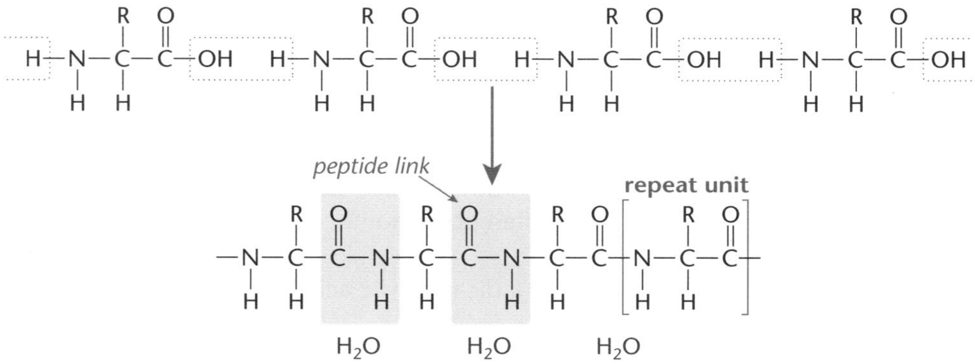

In the **repeat unit** shown here the R group may differ.

Polyamides

Proteins and polypeptides are natural condensation polymers. The link between the amino acid monomer units is usually described as a **peptide link** but chemically this is identical to an **amide** group. Hence polypeptides and proteins are **natural polyamides**.

Nylon, proteins and polypeptides are all polyamides.

Nylon-6,6 was the first man-made condensation polymer and was synthesised as an artificial alternative to natural protein fibres such as wool and silk.

The principle used was to mimic the natural polymerisation process above but, instead of using an amino acid monomer with two different functional groups, **two** chemically different monomers are usually used:

Compare the use of two monomers in the production of synthetic nylon-6,6 with the natural condensation polymerisation using amino acids only.

- a dicarboxylic acid **A** 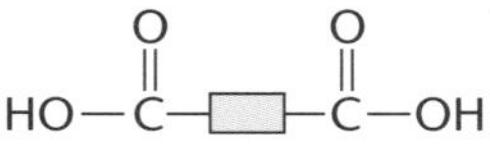
- a diamine **B** 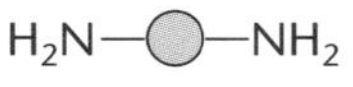

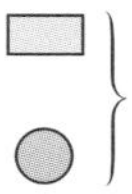

Each diamine molecule bonds to a dicarboxylic acid molecule with loss of water molecule.

Many different polyamides can be made using different carbon chains or rings which bridge the double functional group.

- the two monomers **A** and **B** join alternately: **–A–B–A–B–A–B–A–B–**

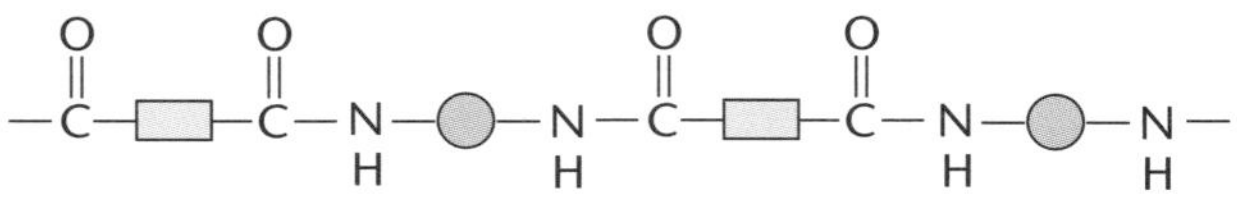

Nylon-6,6 gets its name from the number of carbon atoms in each monomer (diamine first).

The diamine has 6 carbon atoms.

The dicarboxylic acid has 6 carbon atoms.

Hence: nylon-6,6.

The diagrams below show how *nylon-6,6* is formed from its monomers.

hexane-1,6-dioic acid

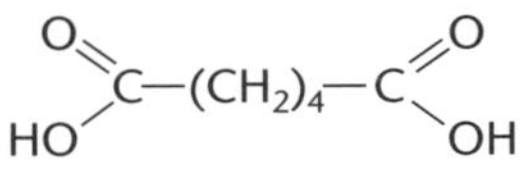

1,6-diaminohexane

$H_2N{-}(CH_2)_6{-}NH_2$

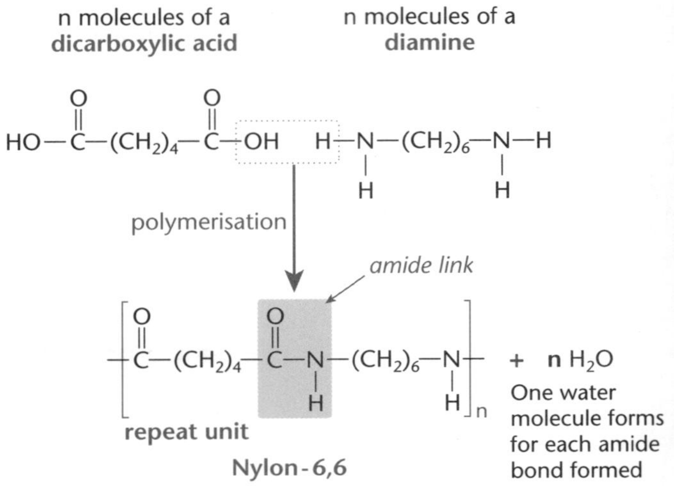

The diacyl chloride for preparing nylon-6,6 would be $ClOC(CH_2)_4COCl$.

Polyamides can also be made using a diacyl chloride instead of a dicarboxylic acid.

- The greater reactivity of an acyl chloride results in easier polymerisation.
- Hydrogen chloride is lost instead of water.

Polyesters

Learn the principle behind condensation polymerisation.

In exams, you may be required to predict structures from unfamiliar monomers.

However, the principle is the same.

Polyesters are polymers made by a condensation reaction between monomers with formation of an ester group as the linkage between the molecules.

The first man-made polyester produced was *Terylene*. As with polyamides, polyesters are used as fibres for clothing.

As with artificial polyamides, man-made polyesters are usually made from **two** different monomers:

- a dicarboxylic acid
- a diol

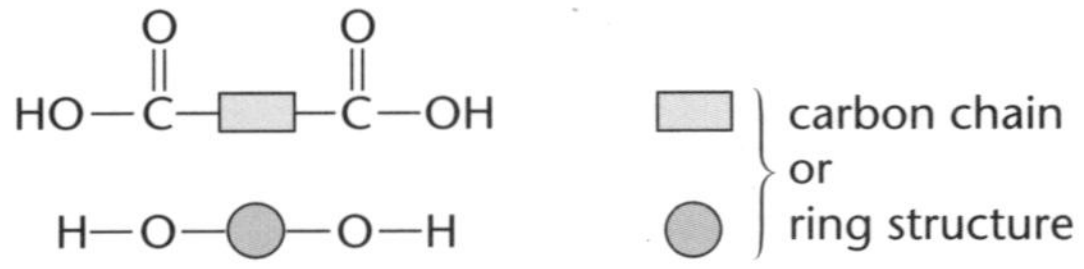

Each diol molecule bonds to a dicarboxylic acid molecule with loss of a water molecule.

Many different polyesters can be made by using different carbon chains or rings bridging the double functional group.

The diagrams below show how *Terylene* is formed from its monomers.

This is the same basic principle as for polyamides.

benzene-1,4-dicarboxylic acid

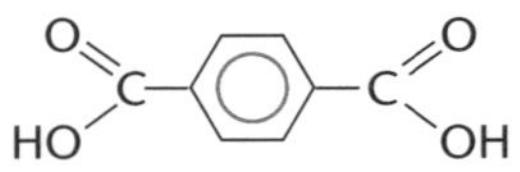

monomer A

ethane-1,2-diol

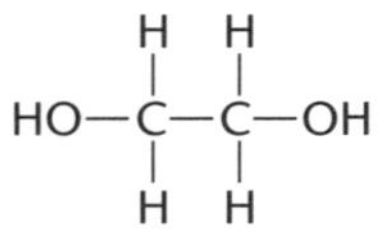

monomer B

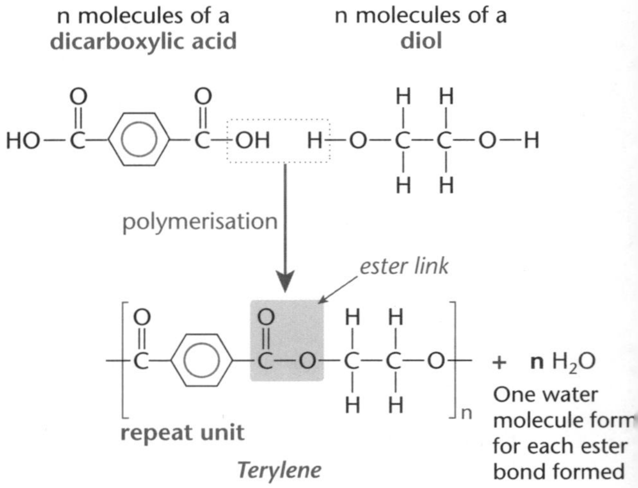

Other polyesters include Kevlar, one of the hardest materials known. Kevlar is used for bulletproof vests, belts for radial tyres, cables and reinforced panels in aircraft and boats.

Disposal of polymers

AQA	M4	SALTERS	M4
EDEXCEL	M5	WJEC	CH4
OCR	M4	NICCEA	M4
NUFFIELD	M6		

Problems with addition polymers

Addition polymers are non-polar and chemically inert.
This creates potential environmental problems during disposal of polymers.
Disposal by landfill causes long-term problems.

- Addition polymers are **non-biodegradable** and take many years to break down.

Disposal by burning can produce toxic fumes.

- Depolymerisation produces poisonous monomers.
- Disposal of poly(chloroethene) (*pvc*) by incineration can lead to the formation of very toxic dioxins if the temperature is too low.

Condensation polymers

Condensation polymers are polar.

- Condensation polymers are broken down naturally by **hydrolysis** into their monomer units. They are therefore biodegradable and easier to dispose off than addition polymers.

Progress check

1 (a) Draw the structure of the monomer needed to make poly(tetrafluoroethene).
(b) Draw the repeat unit of the polymer perspex, made from the monomer methyl 2-methylpropenoate, shown below.

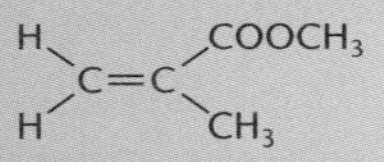

2 (a) Show the structures of the two monomers needed to make nylon-4,6.
(b) Draw a short section of nylon 4,6 and show its repeat unit.

3 Draw the structures of the two monomers needed to make the polyester *kevlar*, shown below.

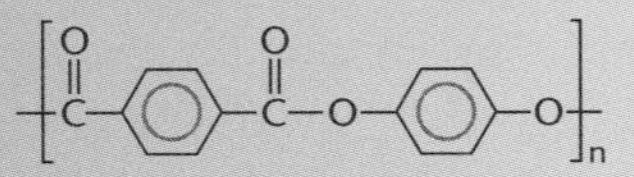

1 (a)
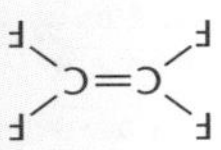
(b)
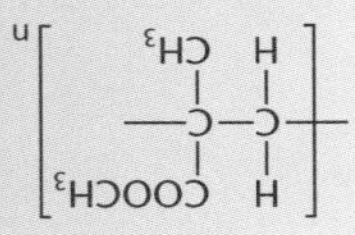

2 (a) $H_2N–(CH_2)_4–NH_2$ and $HOOC–(CH_2)_4–COOH$
(b)
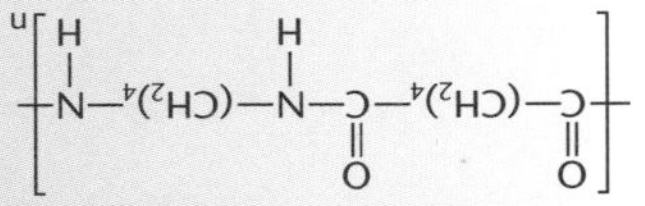

3

Sample question and model answer

The characteristic reaction of benzene is **electrophilic substitution**.

(a) Select a reaction of benzene which illustrates this type of reaction. Give the reagents, the equation, the conditions under which it occurs and the name of the organic product for the reaction you have chosen.

Reagent(s): concentrated HNO_3/H_2SO_4 ✓

Equation: $C_6H_6 + HNO_3 \longrightarrow C_6H_5NO_2 + H_2O$ ✓

Conditions: warm to 55°C ✓

Name of organic product: nitrobenzene ✓ [4]

> This is standard bookwork that must be learnt.
> Other reactions could have been chosen (e.g. bromination and alkylation).
> Remember that 'reagents' are the chemicals 'out of the bottle' that are reacted with the organic compound.
> You must give the full name or formula for any 'reagent'.

(b) For the reaction selected in (a):

(i) identify the electrophile

NO_2^+ ✓

(ii) give an equation to show its formation;

$HNO_3 + H_2SO_4 \longrightarrow H_2NO_3^+ + HSO_4^-$ ✓

$H_2NO_3^+ \longrightarrow NO_2^+ + H_2O$ ✓

> More standard bookwork. An alternative response that would get both marks here is:
> $HNO_3 + 2H_2SO_4 \rightarrow NO_2^+ + 2HSO_4^-$

(iii) give the mechanism for the substitution reaction.

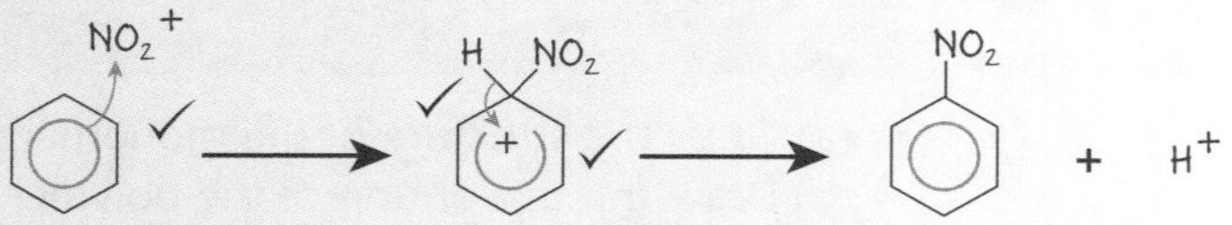

[6]

> Notice how precise you need to be to score all three marks:
> - arrow to electrophile ✓
> - correct intermediate ✓
> - arrow on C–H for loss of H^+ ✓.

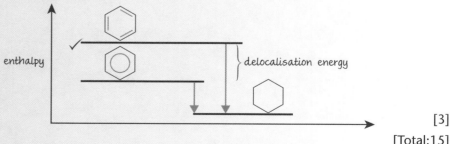

(c) Give two **specific** safety precautions you would need to take in **carrying out** the reaction in (a).

Use of gloves to prevent reagents coming in contact with skin ✓

Fume cupboard to prevent breathing in of poisonous fumes ✓ [2]

> Notice that the answer given here justifies each safety precaution given.

(d) The enthalpy of hydrogenation of a single C–C bond is in the order of $-120\ kJ\ mol^{-1}$.

(i) Assuming that benzene consists of a ring with three separate double bonds, predict the enthalpy change for the reaction.

$3 \times -120\ kJ\ mol^{-1} = -360\ kJ\ mol^{-1}$ ✓

> An easy mark.
> Don't forget the sign!

(ii) The enthalpy of hydrogenation of a benzene is actually $-205\ kJ\ mol^{-1}$. What can you deduce from this and your answer to part (i) about the stability of the benzene ring?

Use an enthalpy level diagram to illustrate your answer.

Benzene has extra stability arising from delocalisation of π-electrons ✓

[3]

[Total:15]

Edexcel Specimen paper Unit Test 4 Q3 January 2000 modified

> An enthalpy diagram is a good way of showing the delocalisation energy of delocalised benzene.
> 1 mark here is awarded for the relative positions of the delocalised and Kekulé structures of benzene.

Practice examination questions

1

A hydrocarbon is known to contain a benzene ring. It has a relative molecular mass of 106 and has the following composition by mass: C, 90.56%; H, 9.44%.

(a) (i) Use the data above to show that the empirical formula is C_4H_5.
(ii) Deduce the molecular formula.
(iii) Draw structures for all possible isomers of this hydrocarbon that contain a benzene ring. [7]

(b) In the presence of a catalyst (such as aluminium chloride), one of these isomers, **A** reacts with chlorine to give only one monochloro-product, **B**.
(i) Deduce which of the isomers in **(a)(iii)** is **A**.
(ii) Draw the structure of **B**. [2]

[Total: 9]

Cambridge Chains and Rings Q2 (a)–(b) Jun 1997

2

(a) Explain why ethylamine is a Brønsted–Lowry base. [2]

(b) Why is phenylamine a weaker base than ethylamine? [2]

(c) Ethylamine can be prepared from the reaction between bromoethane and ammonia.
(i) Name the type of reaction taking place. [1]
(ii) Give the structures of three other organic substitution products which can be obtained from the reaction between bromoethane and ammonia. [3]

(d) Write an equation for the conversion of ethanenitrile into ethylamine and give one reason why this method of synthesis is superior to that in part (c). [2]

[Total: 10]

Assessment and Qualifications Alliance Specimen Test Unit 4 Q6 2000

3

2-Aminopropanoic acid (alanine), $CH_3CH(NH_2)CO_2H$, has a chiral centre and hence can exist as two optical isomers.

(a) (i) State what is meant by a *chiral centre*.
(ii) Explain how a chiral centre gives rise to optical isomerism.
(iii) Draw diagrams to show the relationship between the two optical isomers. State the bond angle around the chiral centre. [5]

(b) In aqueous solution, 2-aminopropanoic acid exists in different forms at different pH values. The zwitterion predominates between pH values of 2.3 and 9.7. Draw the displayed formula of the predominant form of 2-aminopropanoic acid at pH values of 2.0, 6.0 and 10.0. [3]

(c) 2-Aminopropanoic acid can react with an amino acid, **K**, to form the dipeptide below.

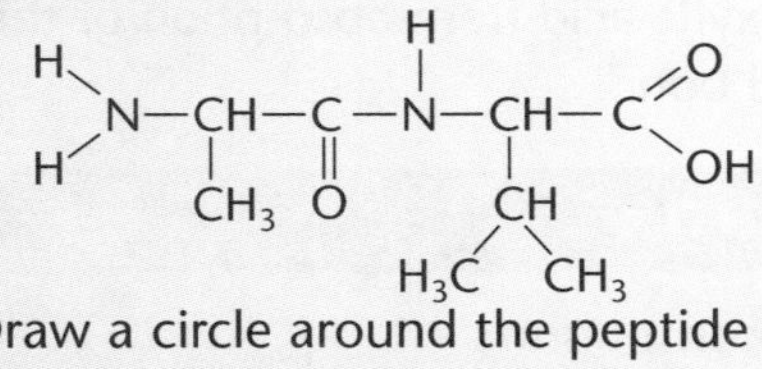

(i) Draw a circle around the peptide linkage.
(ii) Draw the displayed formula of the amino acid, **K**.
(iii) 2-Aminopropanoic acid can react with the amino acid, **K**, to form a different dipeptide from that shown above. Draw the structural formula of this other dipeptide. [3]

[Total: 11]

Cambridge Chains and Rings Q5 June 1998

Chapter 7

Analysis and synthesis

The following topics are covered in this chapter:

- *Infra-red spectroscopy*
- *Mass spectrometry*
- *N.m.r. spectrosopy*
- *Organic synthetic routes*

7.1 Infra-red spectroscopy

After studying this section you should be able to:

- *know that i.r. spectroscopy can be used to identify functional groups*
- *interpret a simple infra-red spectrum*

Using infra-red spectroscopy in analysis

AQA	M4	SALTERS	M5
EDEXCEL	M5	WJEC	CH4
OCR	M4	NICCEA	M5
NUFFIELD	M6		

The frequency of i.r. absorption is measured in *wavenumbers*, units: cm^{-1}.

Basic principles

Molecules are able to convert energy from infra-red radiation into energy to vibrate their bonds. Different bonds absorb different frequencies of infra-red radiation. An infra-red spectrum is obtained by passing a range of infra-red frequencies through a compound producing **absorption peaks**. The frequencies of the absorption peaks can be matched to those of known bonds to identify structural features in an unknown compound.

KEY POINT

Infra-red spectroscopy is useful for identifying the functional groups in a molecule.

Important i.r. absorptions

You don't need to learn the absorption frequencies – the data is provided.

bond	*functional group*	*wavenumber/cm*$^{-1}$
O–H	hydrogen bonded in alcohols	3230 – 3550
N–H	amines	3100 – 3500
O–H	hydrogen bonded in carboxylic acids	2500 – 3300 (broad)
C=O	aldehydes, ketones, carboxylic acids, esters	1680 – 1750
C–O	alcohols, esters	1000 – 1300

An infra-red spectrum is particularly useful for identifying:

- an **alcohol** from absorption of the O–H bond
- a **carbonyl** compound from absorption of the C=O bond
- a **carboxylic** acid from absorption of the C=O bond **and** broad absorption of the O–H bond.

Interpreting infra-red spectra

AQA	M4	SALTERS	M4
EDEXCEL	M5	WJEC	CH4
OCR	M4	NICCEA	M5
NUFFIELD	M6		

Carbonyl compounds (aldehydes and ketones)

Butanone,
$CH_3COCH_2CH_3$

- C=O absorption 1680 to 1750 cm^{-1}

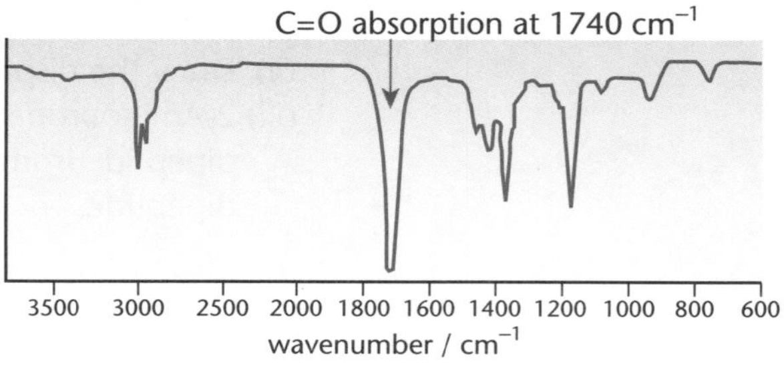

Infra-red spectroscopy is most useful for identifying C=O and O–H bonds.

Look for the distinctive patterns.

Alcohols

Ethanol,
C_2H_5OH

- O–H absorption 3230 to 3500 cm^{-1}
- C–O absorption 1000 to 1300 cm^{-1}

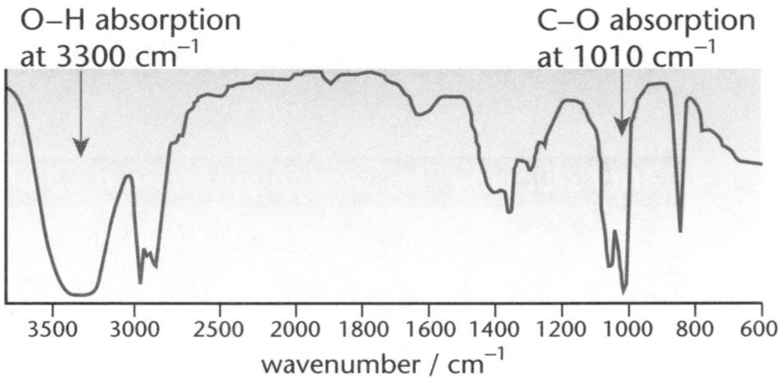

Note that all these molecules contain C–H bonds which absorb in the range 2840 to 3045 cm^{-1}.

Carboxylic acids

Propanoic acid,
C_2H_5COOH

- Very broad O–H absorption 2500 to 3500 cm^{-1}
- C=O absorption 1680 to 1750 cm^{-1}

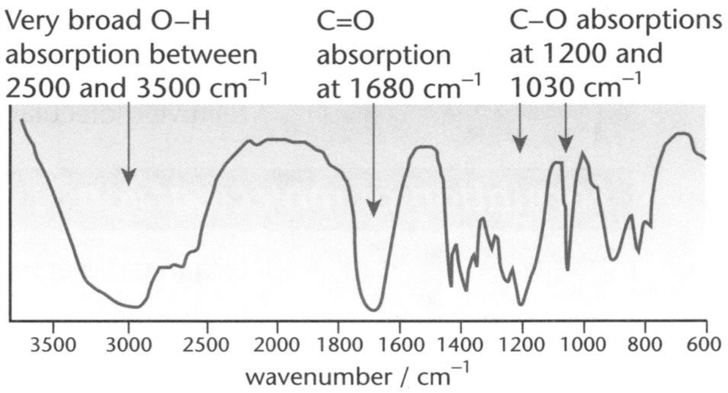

There are other organic groups (e.g. N–H, C=C) that absorb i.r. radiation but the principle of linking the group to the absorption wavenumber is the same.

Ester

Ethyl ethanoate,
$CH_3COOC_2H_5$

- C=O absorption 1680 to 1750 cm^{-1}
- C–O absorption 1000 to 1300 cm^{-1}

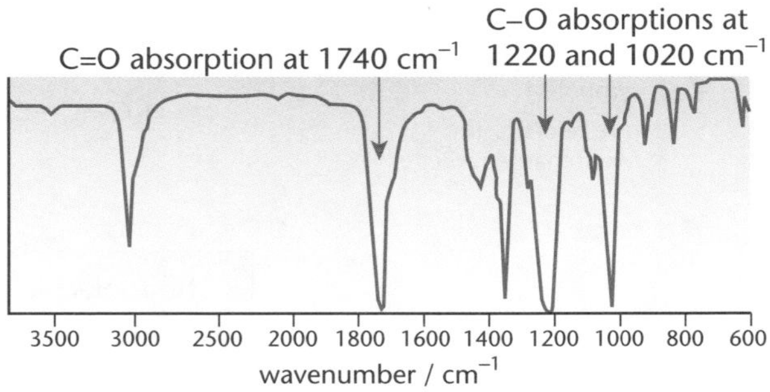

The fingerprint region is unique for a particular compound.

Fingerprint region

- Between 1000 and 1550 cm^{-1}

Many spectra show a complex pattern of absorption in this range.

- This pattern can allow the compound to be identified by comparing its spectrum with spectra of known compounds.

Progress check

1 Ethanol, CH_3CH_2OH was oxidised to ethanal, CH_3CHO and then to ethanoic acid, CH_3COOH. How could you use i.r. spectroscopy to follow the course of this reaction?

2 Two compounds **A** and **B** both have the molecular formula $C_3H_6O_2$. Compound **A** absorbs i.r. at 1720 cm^{-1} and 1030 cm^{-1}. Compound **B** absorbs i.r. at 1700 cm^{-1} and shows broad absorption between 2600 and 3300 cm^{-1}. Identify possible structures for **A** and **B**.

1 Ethanol absorbs at about 3300 cm^{-1} (OH); ethanal absorbs at about 1700 cm^{-1}; ethanoic acid absorbs at about 1700 cm^{-1} and 2500–3500 cm^{-1} (broad)

2 **A**: $HCOOC_2H_5$ or CH_3COOCH_3; **B**: CH_3CH_2COOH.

7.2 Mass spectrometry

After studying this section you should be able to:

- *use a mass spectrum to determine the relative molecular mass of an organic molecule*
- *use the fragmentation pattern of a mass spectrum to deduce a likely structure for an unknown compound*

LEARNING SUMMARY

Key points from AS

- **Measuring relative atomic masses**
  Revise AS page 31

For AS Chemistry, you learnt that mass spectrometry can be used to measure relative atomic masses from a mass spectrum.

For A2 Chemistry, you will study how a mass spectrometer can be used to measure relative molecular masses and identify the molecular structure of organic compounds.

Formation of molecular ions

AQA	M4	SALTERS	M5
EDEXCEL	M5	WJEC	CH4
OCR	M4	NICCEA	M5
NUFFIELD	M6		

Organic molecules can be analysed using mass spectrometry.

In the mass spectrometer, organic molecules are bombarded with electrons. This can lead to the formation of a **molecular ion**.

The equation below shows the formation of a molecular ion from butanone, $CH_3COCH_2CH_3$.

The molecular ion peak is usually give the symbol M.

m/e refers to the mass/charge ratio.

molecular ion, m/e: 72

The molecular ion peak, M, provides the relative molecular mass of the compound.

KEY POINT

Formation of fragment ions

AQA	M4	SALTERS	M5
EDEXCEL	M5	WJEC	CH4
OCR	M5.4	NICCEA	M5
NUFFIELD	M6		

In the conditions within the mass spectrometer, some molecular ions are fragmented by bond fission.

- The bond fission that takes place is a fairly random process:
  - different bonds are broken
  - a **mixture** of **fragment ions** is obtained.
- The mass spectrum contains both the molecular ion and the mixture of fragment ions.

The fragmentation pattern provides clues about the molecular structure of the compound.

Mass spectrometry of organic compounds is useful for identifying:
- the relative molecular mass
- parts of the structural formula.

KEY POINT

Fragmentation of butane, C_4H_{10}

Bond fission forms a fragment ion and a free radical. The diagram below shows that fission of the same C–C bond in a butane molecule can form two different fragment ions.

The mass spectrum of butane would contain peaks for these four ions:
$C_4H_{10}^+$: m/e = 58
$C_3H_7^+$: m/e = 43
$C_2H_5^+$: m/e = 29
CH_3^+: m/e = 15

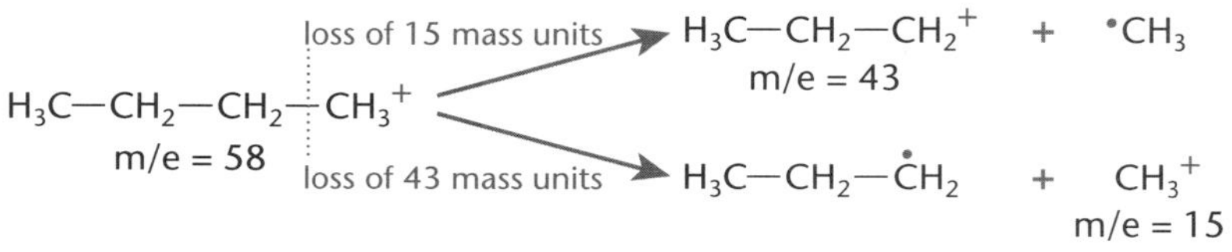

Note that only ions are detected in the mass spectrum.

Uncharged species such as the free radical cannot be deflected in the mass spectrometer.

- Fragmentation of the molecular ion produces a fragment ion by loss of a free radical.
- The mass spectrometer only detects the fragment ion.
- The fragment ion may itself fragment into another ion and free radical.

The mass spectrum of butanone

The mass spectrum of butanone, $CH_3CH_2COCH_3$, is shown below. The m/e values of the main peaks have been labelled.

The M+1 peak is a small peak 1 unit higher than the molecular ion peak.

The origin of the M+1 peak is the small proportion of carbon–13 in the carbon atoms of organic molecules.

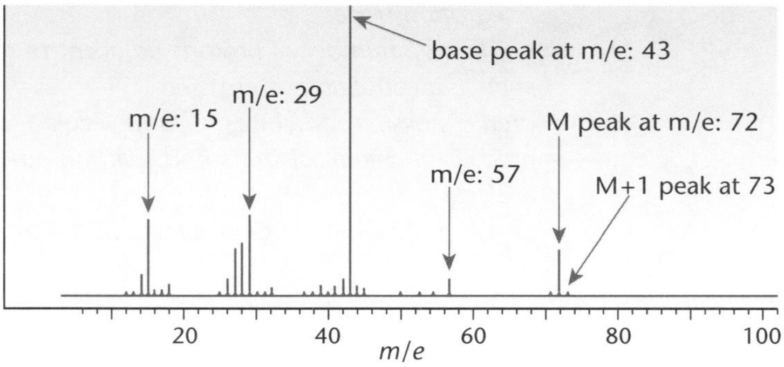

The table below shows the identity of the main peaks.

A common mistake in exams is to show both the fragmentation products as ions.

m/e	*ion*	*fragment lost*
72	$CH_3COCH_2CH_3^+$	-
57	$CH_3CH_2CO^+$	$CH_3\bullet$
43	CH_3CO^+	$CH_3CH_2\bullet$
29	$CH_3CH_2^+$	$CH_3CO\bullet$
15	CH_3^+	$CH_3CH_2CO\bullet$

The most abundant peak in the mass spectrum is the base peak.

The base peak is given a relative abundance of 100. Other peaks are compared with this base peak.

- The base peak in the mass spectrum of butanone has m/e: 43, formed by the loss of a $CH_3CH_2\bullet$ free radical:

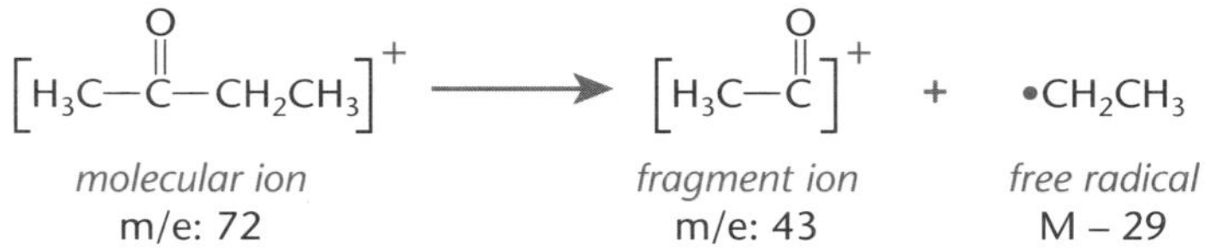

Common patterns in mass spectra

Different fragmentations are possible depending on the structure of the molecular ion.

- The table above shows common fragment ions and fragmented free radicals.
- The skill in interpreting a mass spectrum is in searching for known patterns, evaluating all the evidence and reassembling all the information into a molecular structure.
- You should also look for a peak at m/e 77 or loss of 77 units. This is a giveaway for the presence of a phenyl group, C_6H_5, in a molecule.

Progress check

1 Write an equation for the formation of the fragment ion at m/e=29 in the mass spectrum of butanone above.

2 What peaks would you expect in the mass spectrum of pentane?

1 $CH_3CH_2COCH_3^+ \longrightarrow CH_3CH_2^+ + CH_3CO\bullet$

2 m/e: 72, $C_5H_{12}^+$; m/e: 57, $C_4H_9^+$; m/e: 43, $C_3H_7^+$; m/e: 29, $C_2H_5^+$; m/e CH_3^+, 15.

7.3 N.m.r. spectroscopy

After studying this section you should be able to:

LEARNING SUMMARY

- *predict the different types of proton present in a molecule from chemical shift values*
- *predict the relative numbers of each type of proton present from an integration trace*
- *predict the number of protons adjacent to a given proton from the spin-spin coupling pattern*
- *predict possible structures for a molecule*
- *predict the chemical shifts and splitting patterns of the protons in a given molecule*
- *describe the use of D_2O in n.m.r. spectroscopy*

Nuclear magnetic resonance

AQA	M4	SALTERS	M4, 5
EDEXCEL	M5	WJEC	CH4
OCR	M4	NICCEA	M5
NUFFIELD	M6		

Key principles

The nucleus of an atom of hydrogen (i.e. a proton) has a magnetic spin. When placed in a strong electromagnetic field:

- the nucleus can absorb energy from the **radio-frequency** region of the spectrum to move to a higher energy state
- **nuclear magnetic resonance** occurs as protons resonate between their spin energy states.

Only nuclei with an odd number of nucleons (neutrons + protons) possess a magnetic spin: e.g. ^1H, ^{13}C. Proton n.m.r. spectroscopy is the most useful general purpose technique.

The different nuclear spin states in an applied magnetic field

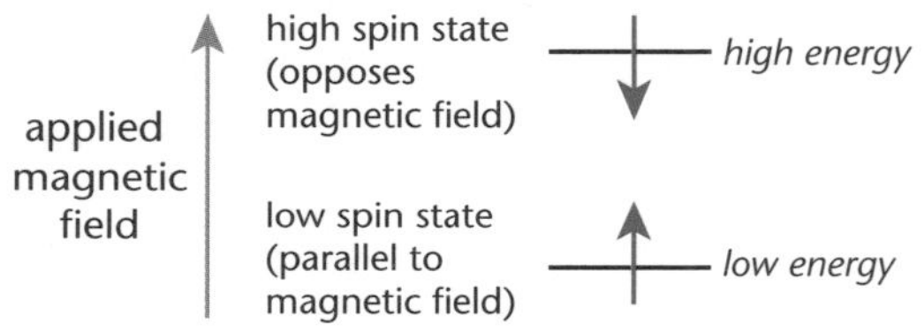

The energy gap is equal to that provided by radio waves.

An n.m.r. spectrum shows **absorption peaks** corresponding to the radio-frequency absorbed.

Chemical shift, δ

Electrons around the nucleus **shield** the nucleus from the applied field.

- The magnetic field at the nucleus of a particular proton is different from the applied magnetic field.

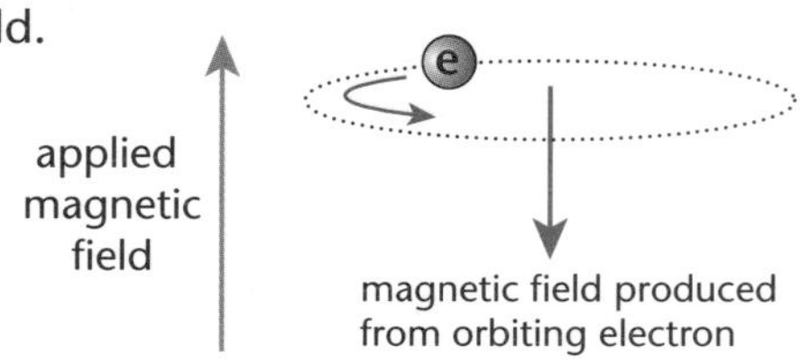

Protons in different environments absorb at different chemical shifts.

- Different radio-frequencies are absorbed, depending on the **environment** of the proton.
- **Chemical shift** is a measure of the magnetic field experienced by protons in different environments resulting from nuclear shielding.

KEY POINT

Chemical shift, δ, is measured relative to a standard: tetramethylsilane (TMS), $Si(CH_3)_4$.
- The chemical shift of TMS is defined as δ = 0.

The frequencies of the absorption peaks in an n.m.r. spectrum can be matched to those of known types of protons. This allows structural features of an unknown compound to be identified.

Typical chemical shifts

Chemical shifts indicate the **types of protons** and **functional groups** present. The table below shows typical chemical shift values for protons in different chemical environments.

The presence of an electronegative atom or group causes chemical shift 'downfield'. This is called 'deshielding'.

Notice the chemical shift caused by the carbonyl group, oxygen, halogens and a benzene ring.

You don't need to learn these chemical shifts – the data is provided on exam papers.

type of proton	*chemical shift, δ*
R-CH_3	0.7–1.6
R-CH_2-R	1.2–1.4
$H_3C-C(=O)-$ R$-CH_2-C(=O)-$	2.0–2.9
$C_6H_5-CH_3$ $C_6H_5-CH_2-R$	2.3–2.7
X-CH_3 X-CH_2-R (X = halogen)	3.2–3.7
–O-CH_3 –O-CH_2-R	3.3–4.3
R–O–**H**	3.5–5.5
C_6H_5–H	7
$H_3C-C(=O)-H$	9.5–10
$H_3C-C(=O)-OH$	11.0–11.7

The only reliable means of identifying O–H in n.m.r. is to use D_2O: see Identifying O–H protons, p. 139.

- The actual chemical shift may be slightly different depending upon the actual environment of the proton.
- The chemical shift for O–**H** can vary considerably and depends upon concentration, solvent and other factors.

Interpreting low resolution n.m.r. spectra

AQA	M4	SALTERS	M4, M5
EDEXCEL	M5	WJEC	CH4
OCR	M4	NICCEA	M5
NUFFIELD	M6		

Low resolution n.m.r. spectrum of ethanol

A low resolution n.m.r. spectrum of ethanol shows absorptions at **three** chemical shifts showing the **three different types of proton**:

The relative areas of each peak are usually measured by running a second n.m.r. spectrum as an *integration trace.*

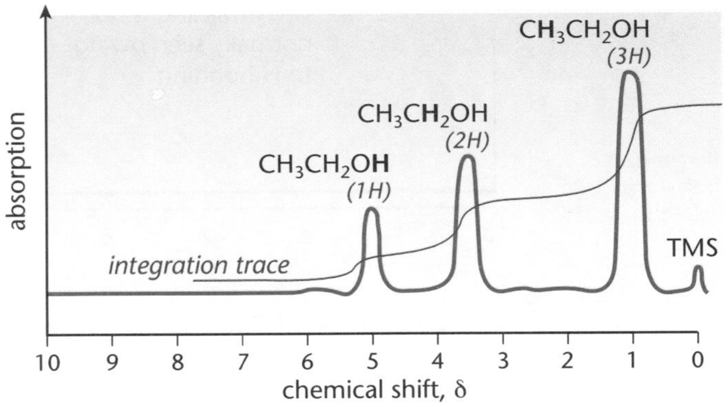

An n.m.r. spectrum is obtained in solution. The solvent must be proton-free, usually CCl_4 or $CDCl_3$ being used.

- The **area** under each peak is in direct proportion to the **number of protons** responsible for the absorption.
- The three chemical shifts can be matched to the table of chemical shifts above so that the protons responsible for each absorption can be identified.

chemical shift	number of protons	environment	type of proton
δ = 1.0	3H	$\mathbf{CH_3}CH_2OH$	CH_3 adjacent to a carbon chain
δ = 3.5	2H	$CH_3\mathbf{CH_2}OH$	CH_2 adjacent to –O
δ = 4.9	1H	$CH_3CH_2\mathbf{OH}$	OH

KEY POINT

A low resolution n.m.r. spectrum is useful for identifying:
- the **number** of different types of proton from the number of peaks
- the **type** of environment of each proton from the chemical shift
- **how many** protons of each type from the integration trace.

Interpreting high resolution n.m.r spectra

AQA	M4	WJEC	CH4
OCR	M4	NICCEA	M5

A high resolution n.m.r. spectrum shows splitting of peaks into a pattern of sub-peaks.

Different types of proton have different chemical shifts.

Spin–spin coupling patterns:
- arise from interactions between protons on **adjacent** carbon atoms which have **different chemical shifts**
- indicate the number of **adjacent** protons.

Equivalent protons (i.e. protons with the same chemical shift) will **not** couple with one another.

The **spin-spin coupling** pattern shows as a **multiplet** – a doublet, triplet, quartet, etc.

A singlet is next to C.
A doublet is next to CH.
A triplet is next to CH_2.
A quadruplet is next to CH_3.

KEY POINT

We can interpret the spin-spin coupling pattern using the **n+1 rule**.
For **n adjacent protons**, the number of peaks in a multiplet = **n+1**.

The high resolution n.m.r. spectrum of ethanol

The high resolution n.m.r. spectrum of ethanol shows spin-spin coupling patterns – some of the signals have been split into multiplets:

The multiplicity (doublet, triplet, quartet) does not indicate the number of protons on that carbon. The number of protons is given by the integration trace.

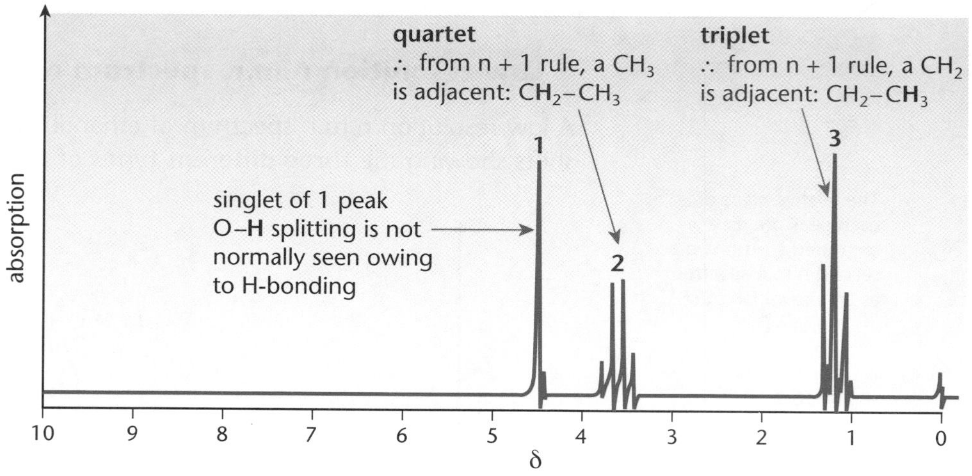

Notice the different sizes of each sub-peak within the splitting pattern:
- doublet 1:1
- triplet 1:2:1
- quartet 1:4:4:1.

The **chemical shift** identifies the **type of protons** responsible for each peak.

The **n+1** rule can be used to identify the **number of adjacent protons**.

chemical shift	*environment*	*number of adjacent protons (n)*	*splitting pattern (n+1)*
δ = 1.2	$\mathbf{CH_3}CH_2OH$	2H	2+1 = 3: **triplet**
δ = 3.6	$CH_3\mathbf{CH_2}OH$	3H	3+1 = 4: **quartet**
δ = 4.5	$CH_3CH_2\mathbf{OH}$	–	**singlet**

- Note that O–**H** splitting is not normally seen owing to H-bonding or exchange with the solvent used.

Equivalent protons do not interact with each other.

- The three equivalent CH_3 protons in ethanol cause splitting of the adjacent CH_2 protons, but not amongst themselves.

> **KEY POINT**
>
> A high resolution n.m.r. spectrum is useful for identifying
> - the **number** of **adjacent** protons from the spin-spin coupling pattern.

Identifying O–H protons

The OH absorption peak at 4.5 δ is absent in D_2O.

O–H protons can absorb at different chemical shifts depending on the solvent used and the concentration. The signals are often broad and usually show no splitting pattern.

These factors can make it difficult to identify O–H protons. However, they can be identified by using deuterium oxide, D_2O.

- A second n.m.r. spectrum is run with a small quantity of D_2O added to the solvent.
- The D_2O exchanges with O–H protons and any such peak will disappear.

Solvents

Deuterated solvents such as $CDCl_3$ are used in n.m.r. spectroscopy. Any protons in a solvent such as $CHCl_3$ would produce a large proton absorption peak. By using $CDCl_3$, this absorption is absent and the spectrum is that of the organic compound alone.

Progress check

1 For each structure, predict the number of peaks in its low resolution n.m.r. spectrum corresponding to the different types of proton and also the number of each different type of proton
(e.g. CH_3CH_2OH has 3 peaks in ratio 3:2:1).
(a) CH_3OH (b) CH_3CH_2CHO (c) CH_3COCH_3 (d) $(CH_3)_2CHOH$.

2 The n.m.r. spectrum of compound **X** (C_2H_4O) has a doublet at δ 2.1 and a quartet at δ 9.8.
(a) Identify compound **X**. (Use the table of chemical shifts on page 137).
(b) How many protons are responsible for each multiplet?

1 (a) 2 peaks, 3:1 (b) 3 peaks 3:2:1 (c) 1 peak
(d) 3 peaks, 6:1:1
2 (a) ethanal, CH_3CHO.
(b) doublet at δ 2.1, 3H; quartet at δ 9.8, 1H.

7.4 Organic synthetic routes

There are many variations possible and the schemes below could not include all reactions without appearing more complicated.

Organic chemists are frequently required to synthesise an organic compound in a multistage process. This is fundamental to the design of new organic compounds such as those needed for modern drugs to combat disease, a new dye or a new fibre. With so many reactions to consider, it is essential to see how different functional groups can be interconverted and summary charts are useful for showing these links. The flow-charts on the next pages show reactions drawn from both the AS and A2 parts of A Level Chemistry.

Construct a scheme of your own using only those reactions in your course.

In revision it is useful to draw your own schemes from memory.

Aliphatic synthetic routes

AQA	M4	SALTERS	M5
EDEXCEL	M5	WJEC	CH4
OCR	M4	NICCEA	M4, M5
NUFFIELD	M6		

The scheme below is based upon two main sets of reactions:

- a set based around bromoalkanes
- a set based around carboxylic acids and their derivatives.

The two sets of reactions are linked via the oxidation of primary alcohols.

- A reaction scheme based upon a secondary alcohol would result in oxidation to a ketone only.

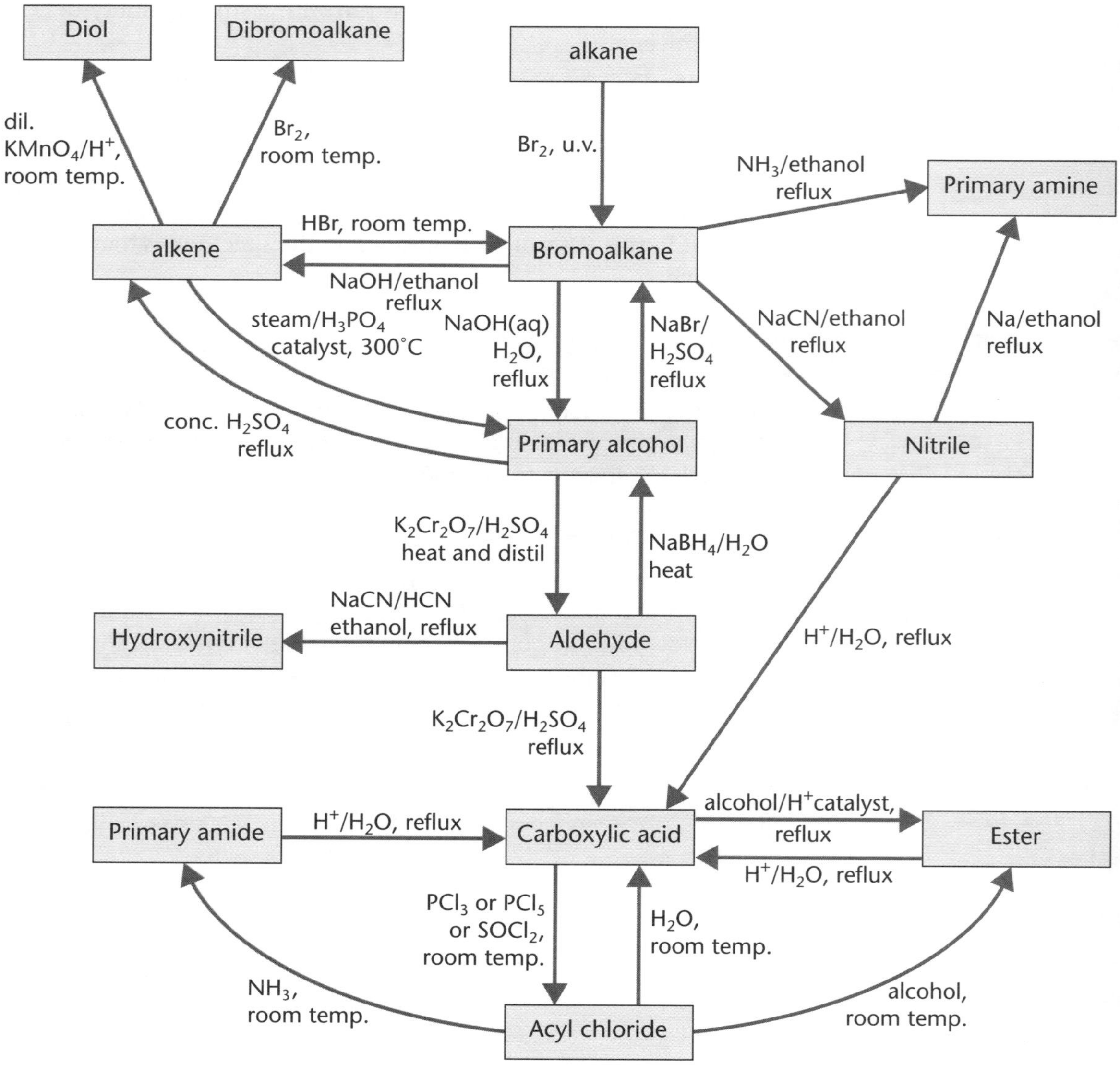

Increasing the carbon chain length

AQA	M4	SALTERS	M4, M5
EDEXCEL	M4, M5	WJEC	CH4
OCR	M4	NICCEA	M4, M5
NUFFIELD	M4, M6		

For more details of these reactions, see 'Carboxylic acids' p. 99 and 'Amines' p. 119.

Hydrogen cyanide is useful for increasing the length of a carbon chain. The nitrile product can easily be reacted further in organic synthesis.

- A nitrile is easily **hydrolysed** by water in hot dilute acid to form a carboxylic acid.
- A nitrile is easily **reduced** by sodium in ethanol to form an amine.

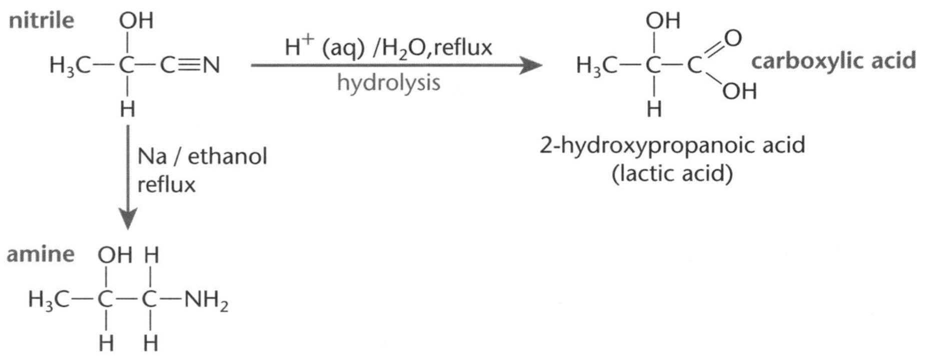

Aromatic synthetic routes

AQA	M4	SALTERS	M4, M5
EDEXCEL	M5	WJEC	CH4
OCR	M4	NICCEA	M5
NUFFIELD	M4, M6		

Compared with aliphatic organic chemistry at A Level, there are comparatively few aromatic reactions.

Benzene

- Reactions involving the benzene ring are mainly electrophilic substitution.

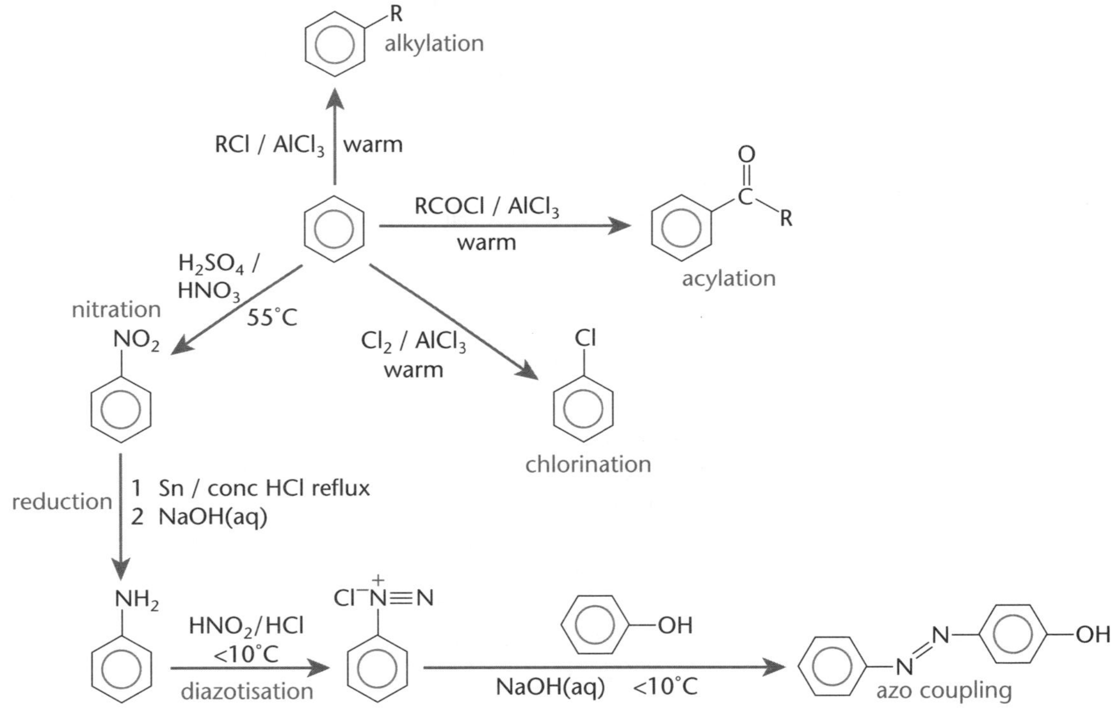

It is essential, if you are to answer such questions, that you thoroughly learn suitable reagents and conditions for all the reactions in your syllabus.

- Questions are often set in exams asking for reagents, conditions or products.
- More searching problems may expect a synthetic route from a starting material to a final product. The synthetic route may include several stages.

Sample question and model answer

An ester **A** of molecular formula $C_4H_8O_2$ was hydrolysed to form a carboxylic acid **B** and an alcohol **C**.

(a) The structures of the four esters of molecular formula $C_4H_8O_2$ are shown below.

$CH_3CH_2COOCH_3$	$CH_3COOCH_2CH_3$	$HCOOCH_2CH_2CH_3$	$HCOOCH(CH_3)_2$
I	II	III	IV

For each structure predict the number of peaks in its low resolution n.m.r. spectrum corresponding to the different types of hydrogen and also the number of each different type of hydrogen.

When we are predicting the number of peaks, we are identifying the number of different types of proton.

In compound IV, $HCOOCH(CH_3)_2$, both the methyl groups (6H) are equivalent and will have the same chemical shift.

Compound I, $CH_3CH_2COOCH_3$, has 3 peaks ✓ in the ratio 3:2:3 ✓
Compound II, $CH_3COOCH_2CH_3$, has 3 peaks ✓ in the ratio 3:2:3 ✓
Compound III, $HCOOCH_2CH_2CH_3$, has 4 peaks ✓ in the ratio 1:2:2:3 ✓
Compound IV, $HCOOCH(CH_3)_2$, has 3 peaks ✓ in the ratio 1:1:6 ✓ [8]

(b) The low resolution n.m.r. spectra of **A** and **B** are shown below with the number of hydrogens for each peak indicated. Use these spectra to decide which of the four esters is the correct one for **A** and explain your answer.

Ester **A**

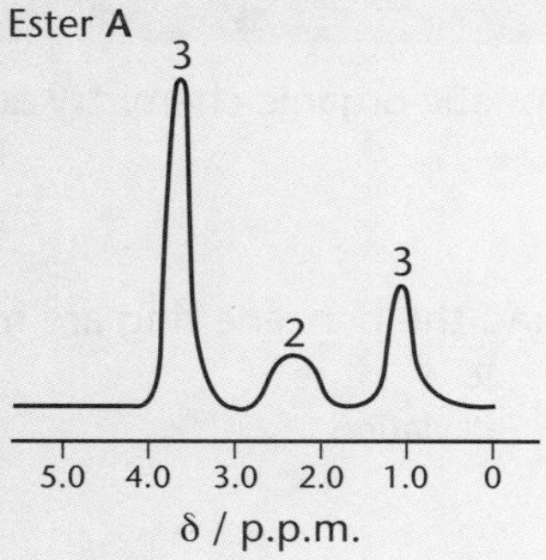

Acid **B**

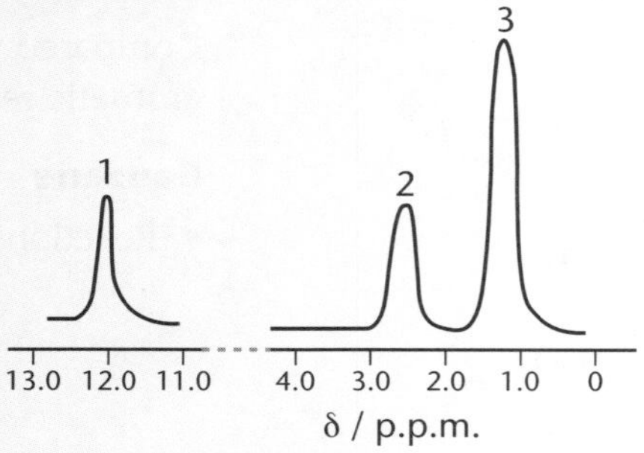

In this question, no chemical shifts were supplied but they were not needed. The problem can be solved using logic and intuition.

Ester A has peaks in the ratio 3:2:3 and must be either I, $CH_3CH_2COOCH_3$, or II, $CH_3COOCH_2CH_3$. ✓
The acid obtained by hydrolysis of I would be CH_3CH_2COOCH – 3 peaks.
The acid obtained by hydrolysis of II would be CH_3COOH – 2 peaks. ✓
Acid B has 3 peaks. ∴ ester A must be I, $CH_3CH_2COOCH_3$. ✓ [3]

(c) Predict, with reasons, what you would expect to see in a high-resolution n.m.r. spectrum of ester **A**.

The n+1 rule is applied here. We have solved this by logic without chemical shift data.

The methyl CH_3 group will be split into a triplet ✓ (2 adjacent protons) ✓
The CH_2 group will be split into a quartet ✓ (3 adjacent protons) ✓
The OCH_3 group will appear as a singlet ✓ (no adjacent protons) ✓
On hydrolysis, the OCH_3 group has been lost. Comparing the spectra of ester A and acid B, this must be the peak at δ 3.7. ✓
∴ the n.m.r. of ester A will show
a singlet at δ 3.7
a quartet at δ 2.2
a triplet at δ 1.0 ✓

[8]

[Total: 19]

NEAB Kinetics and Organic Chemistry Q6(a)-(b) Feb 1995
Part (c) added

Practice examination questions

1

The infra-red spectra shown below were obtained from two isomeric compounds **J** and **K**, of general formula $C_pH_qO_r$.

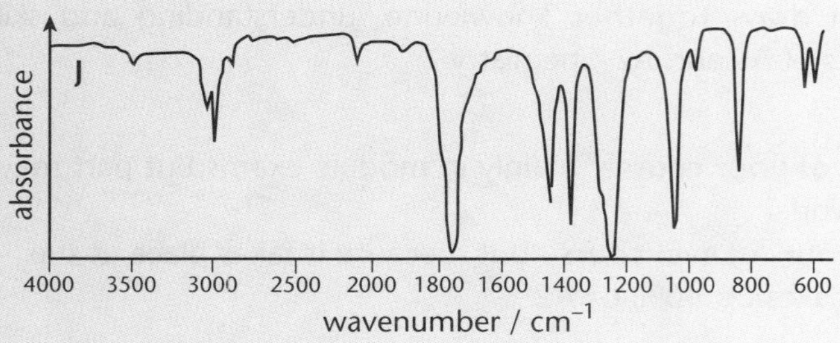

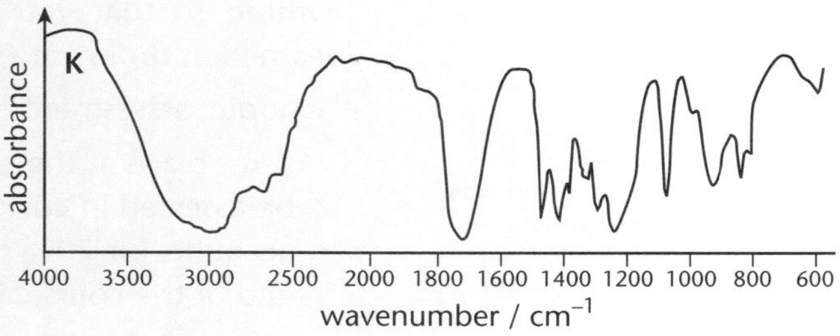

(a) Identify characteristic absorptions and hence name the functional groups present in each of **J** and **K**. (Use the table of i.r. absorptions on page 132.)

(b) Analysis of **J** showed that it contained 54.5% by mass of carbon, and 9.10% by mass of hydrogen. Use these data to determine the empirical formula of **J**.

(c) The M_r of **J** is 88. Show a possible structure for **J**. [10]

[Total: 10]

Cambridge Methods of Analysis and Detection Q6 (d) modified March 2000

2

The proton n.m.r. spectrum of an alcohol, **A**, $C_5H_{12}O$, is shown below.

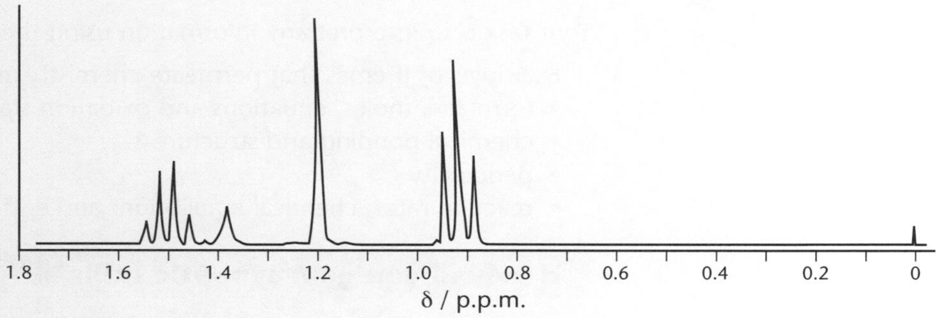

The measured integration trace gives the ratio 0.90 to 0.45 to 2.70 to 1.35 for the peaks at δ 1.52, 1.39, 1.21 and 0.93, respectively.

(a) What compound is responsible for the signal at δ 0? [1]

(b) How many different types of proton are present in compound **A**? [1]

(c) What is the ratio of the numbers of each type of proton? [1]

(d) The peaks at δ 1.52 and δ 0.93 arise from the presence of a single alkyl group. Identify this group and explain the splitting pattern. [5]

(e) What can be deduced from the single peak at δ 1.21 and its integration value? [1]

(f) Give the structure of compound **A**. [1]

[Total: 10]

Assessment and Qualifications Alliance Further Organic Chemistry Q5 March 1999

Chapter 8

Synoptic assessment

What is synoptic assessment?

Part of your chemistry course is assessed using **synoptic questions**. These are written so that you can **draw together** knowledge, understanding and skills learned in **different parts** of AS and A2 Chemistry.

Synoptic assessment emphasises your understanding and your application of the principles included in your Chemistry course.

Synoptic assessment:

- takes place at the **end of your course**, mainly in module exams but part may be assessed in coursework
- accounts for **20% of your A Level course** but, because it takes place at the end of the course, it makes up **40% of A2**.

What type of questions will be asked?

You make the links between different areas of chemistry yourself. You choose and use the context for your answer.

You will need to answer **two** main types of synoptic question.

1 **You make links** and **use connections** between different areas of chemistry. Some examples are given below.

Using examples drawn from different parts of your chemistry course:
- discuss the role of a lone pair in chemistry…
- discuss the chemistry of water…
- compare the common types of chemical bonding…

The context has been made for you. Often this will be a situation or will involve data that you will not have seen before.

2 **You use ideas and skills** which **permeate chemistry**.

This type of question will be more structured and you are unlikely to have as much choice in how you construct your answer.

Your task is to interpret any information using the 'big ideas' of chemistry.

Examples of themes that permeate chemistry are shown below:
- formulae, moles, equations and oxidation states
- chemical bonding and structure
- periodicity
- reaction rates, chemical equilibrium and enthalpy changes.

How will you gain synoptic skills?

A synoptic question may present you with a new piece of information.

You may be expected to calculate formulae from data, write correct formulae, balance equations and perform quantitative calculations using the Mole concept, use knowledge of the Periodic Table to predict reactions of unfamiliar elements or compounds…

Luckily chemistry is very much a synoptic subject. Throughout your study of chemistry, you apply many of the ideas and skills learnt during AS chemistry or your GCSE course. In studying A2 Chemistry, you will have been using synoptic skills naturally, probably without realising it. You certainly can make little real progress in chemistry without a sound grasp of concepts such as equations, the mole, structure and bonding!

How are synoptic questions different?

Different Exam Boards have used different strategies for their synoptic assessment and some of these are listed below.

- Structured questions: short and extended answers based on important chemical themes drawn from AS and A2 Chemistry.
- An essay-type question.
- A problem set in a practical context.
- Objective questions: multiple-choice and either matching pairs or multiple completion items.

Worked synoptic questions and model answers

Here some of the big ideas of chemistry are being tested: formulae, moles and equations. These will always appear on an exam paper with synoptic questions.

You have nothing to fear from this type of question provided that your chemistry is sound.

Structured type question

1

Nitrogen oxides such as nitrogen dioxide, NO_2, are pollutants and cause considerable harm to the environment.

(a) Nitrogen dioxide and dinitrogen tetroxide exist in the following equilibrium:

$$2NO_2(g) \rightleftharpoons N_2O_4(g)$$

When 11.04 g of nitrogen dioxide were placed in a vessel at a fixed temperature, 5.52 g of dinitrogen tetroxide were produced at equilibrium under a pressure of 100 kPa.

(i) Calculate the mole fractions of NO_2 and N_2O_4 present in the equilibrium mixture.

Equilibrium masses: $NO_2 = 5.52$ g; $N_2O_4 = 11.04 - 5.52 = 5.52$ g.

moles = $\dfrac{\text{mass}}{\text{molar mass}}$.

amount $NO_2 = \dfrac{5.52}{46.0} = 0.120$ mol ✓; amount $N_2O_4 = \dfrac{5.52}{92.0} = 0.060$ mol ✓

When working with mole fractions, remember the total number of moles $0.120 + 0.060 = 0.180$ mol.

mole fraction $(NO_2) = \dfrac{\text{moles}(NO_2)}{\text{total number of moles}} = \dfrac{0.120}{0.180} = 0.667$ ✓

mole fraction $(N_2O_4) = \dfrac{0.060}{0.180} = 0.333$ ✓ [4]

(ii) Calculate the value of the equilibrium constant, K_p, and state its units.

For K_p, don't use square brackets – these mean concentrations!

units: $\dfrac{(\text{kPa})}{(\text{kPa})^2} = \text{kPa}^{-1}$

partial pressure, p = mole fraction x total pressure, P

$p_{NO_2} = 0.667 \times 100 = 66.7$ kPa; $p_{N_2O_4} = 0.333 \times 100 = 33.3$ kPa ✓

$K_p = \dfrac{p_{N_2O_4}}{p_{NO_2}^{\,2}}$ ✓ $= \dfrac{(33.3)}{(66.7)^2} = 7.49 \times 10^{-3}$ ✓ kPa^{-1} ✓ [4]

(iii) State and explain the effect on the mole fraction of NO_2 when the pressure is increased at constant temperature.

Use le Chatelier's process. There are 3 key points here for the 3 marks.

From the equation: $2NO_2(g) \rightleftharpoons N_2O_4(g)$,
the system will oppose an increase in pressure ✓ by moving towards the side of the equation with fewer gas moles. ✓
∴ the $[NO_2]$ decreases and mole fraction of NO_2 decreases. ✓ [3]

(b) State **two** environmental consequences of nitrogen oxides.

nitrogen oxides cause **acid** (rain) ✓, greenhouse effect ✓, photochemical smog/ozone build-up ✓ [2 max]

(c) Not all nitrogen compounds are harmful: some, such as nitrogen fertilisers, are beneficial to man.

A nitrogen fertiliser, **D**, has the composition by mass Na, 27.1%; N, 16.5%; O, 56.4%. On heating, 3.40 g of **D** was broken down into sodium nitrite, $NaNO_2$, and oxygen gas.

Suggest an identity for the fertiliser, **D**, and calculate the volume of oxygen that was formed. [Assume that 1 mole of gas molecules occupy 24 dm^3.]

D Na: N: O = 27.1/23 : 16.5/14.0 : 56.4/16.0 ✓ = $NaNO_3$ ✓

$2NaNO_3 \longrightarrow 2NaNO_2 + O_2$

∴ 0.04 mol ⟶ 0.02 mol ✓

volume of $O_2(g) = 24 \times 0.0200\ dm^3 = 0.48\ dm^3$ ✓ [4]

[Total: 17]

Worked synoptic questions and model answers (continued)

Essay type question

2

Write an essay on the chemistry of ammonia.

In your answer you should include a discussion of the bonding in, and structure of the molecule, consider its industrial preparation and you should bring together its reactions in both inorganic and organic chemistry.

You have been given a little help about what needs to go into your essay.

You will always score more marks by answering 'a bit about everything' than 'a lot about one part'.

Bonding and structure of NH_3.

Ammonia, NH_3 is bonded by 3 covalent bonds ✓. There are 4 centres of electron density around the central N atom (3 covalent bonds + 1 lone pair ✓ of electrons). This gives a pyramidal shape ✓ with a H–N–H bond angle of 107° ✓:

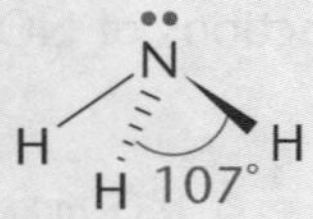

[4]

Industrial preparation of NH_3.

Ammonia is prepared in the Haber process by the reaction of nitrogen with hydrogen, catalysed by iron✓:

$$N_2(g) + 3H_2(g) \rightleftharpoons 2NH_3(g)$$ ✓

The optimum equilibrium yield is a high pressure and low temperature ✓. However, the reaction proceeds too slowly at low temperatures and a compromise is necessary between a high enough temperature to overcome activation energy and as low a temperature as possible to secure a reasonable equilibrium yield. ✓ [4]

The choice of reaction is yours – you make the links.

Ammonia is a Brønsted–Lowry base ✓

e.g: ammonia forms salts with inorganic acids:

$NH_3 + HCl \longrightarrow NH_4^+Cl^-$ ✓

ammonia forms salts with organic acids:

$NH_3 + CH_3COOH \longrightarrow CH_3COO^-NH_4^+$ ✓

To gain full marks in this part you must include reactions from both inorganic and organic reactions of NH_3.

In the marking scheme, there will be a 'ceiling' of marks for each branch of chemistry and for each 'type of reaction'.

A description of ten ligand substitution reactions would only answer from one area of chemistry and, even if correct, would score poorly.

Ammonia reacts with complex aqua-ions,

as a base forming a metal hydroxide precipitate: ✓

$[Cu(H_2O)_6]^{2+}(aq) + 2NH_3(aq) \longrightarrow Cu(OH)_2(H_2O)_4(s) + 2NH_4^+(aq)$ ✓

as a ligand, forming complex ions by ligand substitution with metal aqua-ions: ✓

$[Cr(H_2O)_6]^{3+} + 6NH_3 \longrightarrow [Cr(NH_3)_6]^{3+} + 6H_2O$ ✓

Ammonia can behave as a nucleophile, ✓

reacting with halogenoalkanes in a nucleophilic substitution reaction:✓

$RBr + 2NH_3 \longrightarrow RNH_2 + NH_4^+Br^-$ ✓ [7 max]

[Total: 17]

Assessment and Qualifications Alliance Specimen Unit Test 5 Q8 2000

Practice examination questions

Structured type questions

1

A solution of cobalt(II) chloride was reacted with ammonia and ammonium chloride while a current of air was blown through the mixture. A red compound **X** was produced which contained a complex ion of cobalt. The compound had the following composition.

	% by mass
Co	23.6
N	27.9
H	6.0
Cl	42.5

(a) Use the data to calculate the empirical formula of the compound **X**. [3]

(b) In the reaction in which **X** is formed from cobalt(II) chloride, explain the role of:
  (i) the air
  (ii) the ammonia and ammonium chloride. [3]

(c) A solution of cobalt(II) chloride reacts with concentrated hydrochloric acid to form a stable complex ion. A solution of calcium chloride does not form a corresponding complex. Use your knowledge of atomic structure to explain this difference, and suggest other differences you would expect between the chemistry of cobalt and calcium. [4]

[Total: 10]

Nuffield Specimen Unit Test 6 Q3 2000

2

The structure of the silkworm moth sex attractant, bombykol, is:

$$H_3C–(CH_2)_2–CH=CH–CH=CH–(CH_2)_8–CH_2OH$$

(a) Predict some of the properties you would expect for bombykol. You should comment on:
  (i) the likely solubility of bombykol in water
  (ii) the number of possible geometric isomers
  (iii) its likely reaction with four reactants of your choice.
  Write equations or reaction schemes for the reactions you choose, showing the structures of the organic products clearly. [8]

(b) When bombykol is treated with ozone in the presence of water the molecule splits into fragments wherever there is a C=C double bond.

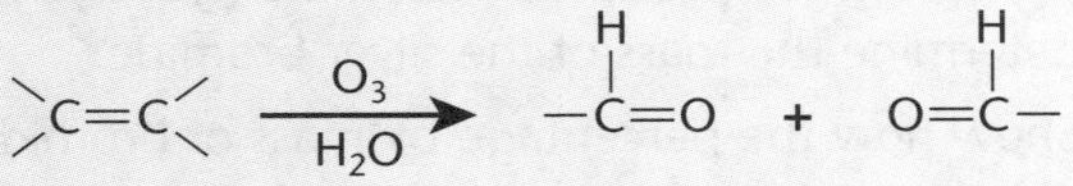

The aldehyde groups are converted to carboxylic acids in oxidizing conditions.

Predict the oxidation products formed when bombykol is reacted with ozone in oxidising conditions. [2]

[Total: 10]

Nuffield Specimen Unit Test 6 Q2 2000

Practice examination questions (continued)

Essay type question

3

Redox reactions are an important type of reaction in chemistry.

Explain what is meant by a redox reaction. Illustrate your answer with **two** examples drawn from inorganic chemistry (one of which should involve a transition element) and **two** examples from organic chemistry. [Total: 14]

Cambridge Trend and Patterns Specimen Unit Test Q4 2000

Interpretation of data

4

Using knowledge, principles and concepts from different areas of chemistry, explain and interpret, as fully as you can, the data given in the table below. In order to gain full credit, you will need to consider each type of information separately and also to link this information together.

compound	*boiling point /K*	*properties of a 0.1 mol dm^{-3} solution*	
		electrical conductivity	*[H$^+$]/mol dm^{-3}*
NaCl	1686	good	1.0×10^{-7}
CH_3COOH	391	slight	1.3×10^{-3}
CH_3CH_2OH	352	poor	1.0×10^{-7}
$AlCl_3$	451	good	3.0×10^{-1}

[Total: 14]

Cambridge Unifying concepts Specimen Unit Test Q4 2000

Practical-based problem

5

You are required to plan an experiment to determine the percentage by mass of bromine in a bromoalkane. The bromoalkane, which boils at 75°C, can be hydrolysed completely by heating with an appropriate amount of boiling, aqueous sodium hydroxide for about 40 minutes. The bromide ion released can be estimated by converting it into a silver bromide precipitate which is subsequently weighed.

(a) Write equations for the reactions which occur.

(b) Describe how you would carry out the hydrolysis, giving details of the apparatus and the conditions which you would use.

(c) Describe, giving details of the apparatus and reagents, how you would obtain a silver bromide precipitate from the hydrolysis solution and how you would determine the mass of the silver bromide.

(d) Show how the percentage by mass of bromide ion, in the original haloalkane, can be calculated.

[Total: 15]

Assessment and Qualifications Alliance Specimen Unit Test 5 Q9 2000

Practice examination answers

1 Reaction rates

1 (a) The colour changes from colourless to orange-brown as iodine forms. ✓ This can be monitored using a colorimeter. ✓ [2]

(b) (i) 1st order with respect to H_2O_2 ✓
Double concentration of H_2O_2, rate doubles ✓
1st order with respect to I^- ✓
Quadruple concentration of I^-, rate x4 ✓
Zero order with respect to H^+ ✓
Double concentration of H^+, rate stays constant ✓ [6]

(ii) rate = $k[H_2O_2][I^-]$ ✓ [1]

(iii) 2.8×10^{-2} $dm^3\ mol^{-1}\ s^{-1}$ ✓✓ (from experiment 1) [2]

(c) (i) The slowest step of a multi-step process ✓ [1]

(ii) $H_2O_2 + I^- \longrightarrow H_2O + IO^-$ ✓✓ [2]

[Total: 14]

2 (a) (i) rate = $k[X][Y]^2$ ✓✓ [2]

(ii) 3 ✓ [1]

(iii) 8 ✓ [1]

(b) (i) Overall order of reaction = 2 ✓
Concentrations of both A and B doubled, rate × 4 ✓ [2]

(ii) Order with respect to B = 0 ✓
When [A] doubled and [B] is constant (experiments 2 and 3), rate × 4.
∴ [B] has no effect ✓ [2]

(iii) rate = $k[A]^2$ ✓ [1]

(iv) $k = 8.75 \times 10^{-3}$ $dm^3\ mol^{-1}\ s^{-1}$ ✓✓ [2]

[Total: 11]

2 Chemical equilibrium

1 (a) (i) $$K_c = \frac{[CH_3COOCH_2CH_3]\,[H_2O]}{[CH_3COOH]\,[C_2H_5OH]}$$ ✓ [1]

(ii) 3.86 ✓✓✓ [3]

(iii) Equilibrium moves to the right ✓ to counteract the added CH_3COOH ✓ [2]

(b) No effect ✓ The catalyst is not in the expression for K_c ✓ [2]

[Total: 8]

2 (a) Yield of SO_3 is less. ✓ Forward reaction is exothermic. ✓ Equilibrium position moves to the left to counteract the increase in temperature. ✓ [3]

(b) Partial pressure of O_2 = 48 kPa. ✓ Mole fraction of O_2 = 48/120 = 0.4 ✓✓ [3]

(c) (i) $K_p = \frac{p_{SO_3}{}^2}{p_{SO_2}{}^2 \times p_{O_2}}$ ✓ [1]

(ii) 2.91×10^{-2} kPa^{-1} ✓✓✓ [3]

[Total: 10]

3 (a) Equilibrium mixture becomes more concentrated ✓ [1]

(b) Increase in pressure. ✓ Equilibrium position moves to the left (the side with least gas molecules) to relieve the increase in pressure. ✓ [2]

(c) $K_p = \frac{p_{NO_2}{}^2}{p_{N_2O_4}}$ ✓ [1]

(d) 2.68 atm ✓✓✓ [3]

[Total: 7]

4 (a) Proton donor ✓ [1]

(b) $HCl(g) + H_2O(l) \rightleftharpoons H_3O^+(aq) + Cl^-(aq)$ ✓
H_2O is a base because it accepts a proton. ✓ [2]

(c) $NH_3(g) + H_2O(l) \rightleftharpoons NH_4^+(aq) + OH^-(aq)$ ✓
H_2O is an acid because it donates a proton. ✓ [2]

(d) $H_2SO_4 + HNO_3 \rightleftharpoons H_2NO_3^+ + HSO_4^-$ ✓
$H_2NO_3^+ \longrightarrow NO_2^+ + H_2O$ ✓
HNO_3 is a base because it accepts a proton. ✓ [3]

(e) (i) Species that only partially dissociates to donate protons✓ [1]

(ii) $K_a = \frac{[H^+(aq)]\,[A^-(aq)]}{[HA(aq)]}$ ✓ [1]

(iii) Concentration of undissociated HX is very small ✓ [1]

(iv) Strong acid ✓ because [HX] is much smaller than $[H^+]$ and $[Cl^-]$. ✓ [2]

[Total: 13]

5 (a) (i) $pH = -\log[H^+(aq)]$ ✓ [1]

(ii) An acid that is partially ionised ✓ [1]
$HCOOH + H_2O \rightleftharpoons HCOO^- + H_3O^+$ ✓ [1]

(b) (i) pH = 0.82 ✓✓ [2]

(ii) pH = 13.9 ✓✓✓ [3]

(iii) pH = 2.28 ✓✓✓✓ [4]

(c) (i) It resists changes in pH ✓ with small amounts of acid or base. ✓ [2]

(ii) pH = 5.26 ✓✓✓ [3]

[Total: 17]

3 Energy changes in chemistry

1 (a)

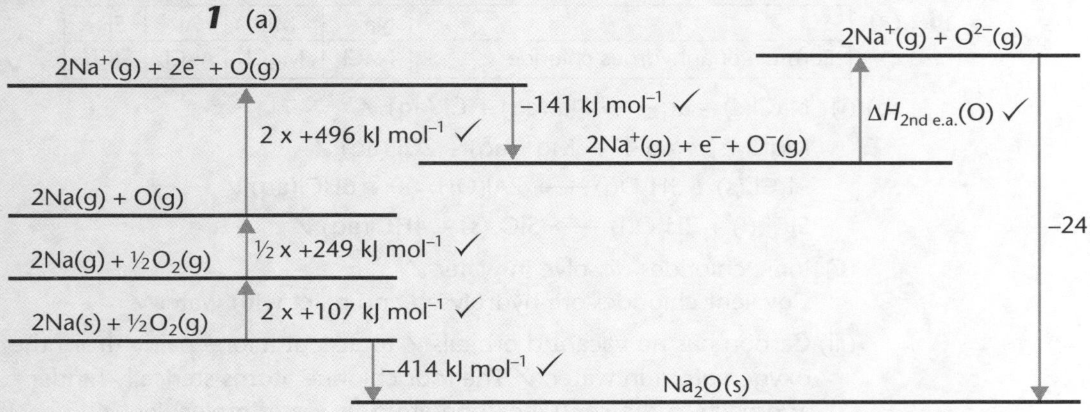

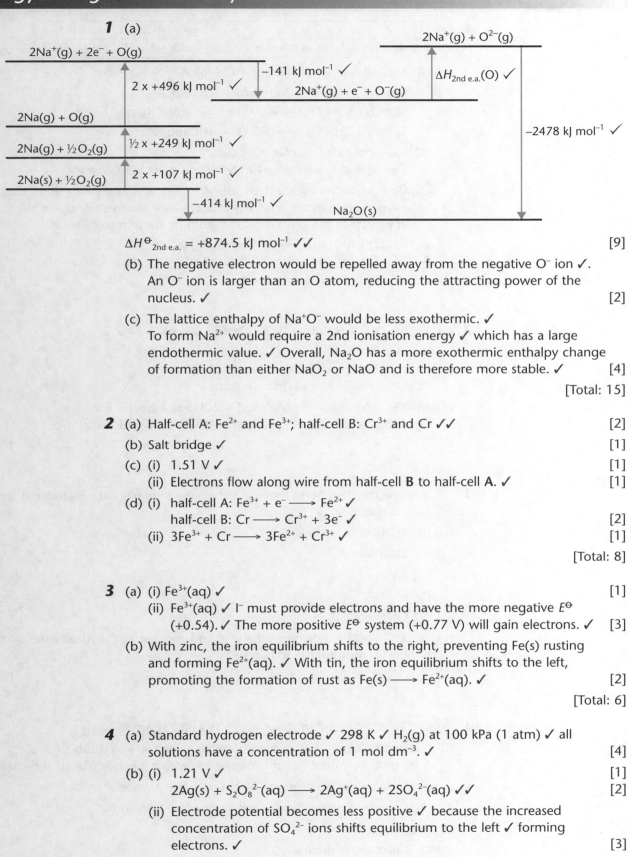

$\Delta H^{\ominus}_{2nd\ e.a.}$ = +874.5 kJ mol^{-1} ✓✓ [9]

(b) The negative electron would be repelled away from the negative O^- ion ✓. An O^- ion is larger than an O atom, reducing the attracting power of the nucleus. ✓ [2]

(c) The lattice enthalpy of Na^+O^- would be less exothermic. ✓ To form Na^{2+} would require a 2nd ionisation energy ✓ which has a large endothermic value. ✓ Overall, Na_2O has a more exothermic enthalpy change of formation than either NaO_2 or NaO and is therefore more stable. ✓ [4]

[Total: 15]

2 (a) Half-cell A: Fe^{2+} and Fe^{3+}; half-cell B: Cr^{3+} and Cr ✓✓ [2]

(b) Salt bridge ✓ [1]

(c) (i) 1.51 V ✓ [1]

(ii) Electrons flow along wire from half-cell **B** to half-cell **A**. ✓ [1]

(d) (i) half-cell A: $Fe^{3+} + e^- \longrightarrow Fe^{2+}$ ✓

half-cell B: $Cr \longrightarrow Cr^{3+} + 3e^-$ ✓ [2]

(ii) $3Fe^{3+} + Cr \longrightarrow 3Fe^{2+} + Cr^{3+}$ ✓ [1]

[Total: 8]

3 (a) (i) $Fe^{3+}(aq)$ ✓ [1]

(ii) $Fe^{3+}(aq)$ ✓ I^- must provide electrons and have the more negative $E^{\ominus}$ (+0.54). ✓ The more positive $E^{\ominus}$ system (+0.77 V) will gain electrons. ✓ [3]

(b) With zinc, the iron equilibrium shifts to the right, preventing Fe(s) rusting and forming $Fe^{2+}(aq)$. ✓ With tin, the iron equilibrium shifts to the left, promoting the formation of rust as $Fe(s) \longrightarrow Fe^{2+}(aq)$. ✓ [2]

[Total: 6]

4 (a) Standard hydrogen electrode ✓ 298 K ✓ $H_2(g)$ at 100 kPa (1 atm) ✓ all solutions have a concentration of 1 mol dm^{-3}. ✓ [4]

(b) (i) 1.21 V ✓ [1]

$2Ag(s) + S_2O_8^{2-}(aq) \longrightarrow 2Ag^+(aq) + 2SO_4^{2-}(aq)$ ✓✓ [2]

(ii) Electrode potential becomes less positive ✓ because the increased concentration of SO_4^{2-} ions shifts equilibrium to the left ✓ forming electrons. ✓ [3]

[Total: 10]

4 The Periodic Table

1 (a)

	Na	Mg	Al	Si
Formula of anhydrous chloride	$NaCl$	$MgCl_2$	Al_2Cl_6	$SiCl_4$

✓✓ [2]

(b) (i) $NaCl(s) + aq \longrightarrow Na^+(aq) + Cl^-(aq)$ ✓
$MgCl_2(s) + aq \longrightarrow Mg^{2+}(aq) + 2Cl^-(aq)$ ✓
$Al_2Cl_6(s) + 3H_2O(l) \longrightarrow 2Al(OH)_3(s) + 6HCl(aq)$ ✓
$SiCl_4(l) + 2H_2O(l) \longrightarrow SiO_2(s) + 4HCl(aq)$ ✓ [4]

(ii) Ionic chlorides dissolve in water ✓
Covalent chlorides are hydrolysed and react with water ✓ [2]

(iii) Carbon has no vacant d orbitals ✓ to accept a lone pair ✓ from the oxygen atom in water. ✓ The four chlorine atoms sterically hinder approach to the central carbon atom by water molecules. ✓ [4]

(c) $Al(OH)_3(s) + 3H^+(aq) \longrightarrow Al^{3+}(aq) + 3H_2O(l)$ ✓
$Al(OH)_3(s) + OH^-(aq) \longrightarrow Al(OH)_4^-(aq)$ ✓ [2]

[Total: 14]

2 (a) (i) $Na_2O(s) + H_2O(l) \longrightarrow 2NaOH(aq)$ ✓ ; pH = 14 ✓
(ii) $SO_2(g) + H_2O(l) \longrightarrow H_2SO_3(aq)$ ✓ ; pH = 3 ✓ [4]

(b) Ionic oxides form alkaline solutions ✓
Covalent oxides form acidic solutions ✓ [2]

(c) (i) $MgCl_2(s) + aq \longrightarrow Mg^{2+}(aq) + 2Cl^-(aq)$ ✓ ; pH = 6 ✓
(ii) $SiCl_4(l) + 2H_2O(l) \longrightarrow SiO_2(s) + 4HCl(aq)$ ✓ ; pH = 1 ✓ [4]

[Total: 10]

3 (a) A dative covalent bond forms ✓ between a lone pair on a ligand and the central metal ion. ✓ [2]

(b) (i) 1,2-diaminoethane ✓ [1]
(ii) Bidentate ✓ [1]
(iii) +3 ✓ [1]
(iv) 6 ✓ [1]
(v) Octahedral ✓ [1]

(c) To $CoCl_2$, add 1,2-diaminoethane ✓ and concentrated hydrochloric acid ✓
Bubble O_2 through the solution ✓. [3]

[Total: 10]

4 (a) Co: $1s^22s^22s^63s^23p^63d^74s^2$ ✓; Co^{2+}: $1s^22s^22s^63s^23p^63d^7$ ✓ [2]

(b) A d-block element has its highest energy electron in a d sub-shell. ✓
A transition element has at least one ion with a partially filled d sub-shell. ✓ [2]

(c) (i) Lone pair of electrons on electronegative oxygen atom ✓ [1]
(ii) Covalent bonding and ✓ dative covalent bonding ✓ [2]
(iii) Octahedral ✓ [1]

(d) (i) Cobalt(II) hydroxide ✓
$[Co(H_2O)_6]^{2+}(aq) + 2OH^-(aq) \longrightarrow Co(OH)_2(H_2O)_4(s) + 2H_2O(l)$ ✓ [2]
(ii) Precipitation ✓ [1]

(iii) $[Co(OH)_6]^{4-}$ ✓ [1]

(iv) the hexaamminecobalt(II) ion ✓ [1]

(v) $[Co(H_2O)_6]^{2+}(aq) + 4Cl^-(aq) \longrightarrow [CoCl_4]^{2-}(aq) + 6H_2O(l)$ ✓
ligand substitution ✓ [2]

[Total: 15]

5 (a) Reducing agent ✓ [1]

(b) Dilute sulphuric acid ✓ [1]

(c) Colourless ⟶ pink ✓ [1]

(d) 0.0362 mol dm^{-3} ✓✓✓✓ [4]

[Total: 7]

5 Isomerism, aldehydes, ketones and carboxylic acids

1 (a) C_3H_6O ✓✓ [2]

(b)

[2]

(c) 2,4-dinitrophenylhydrazine ✓ [2]
Observation: yellow/orange precipitate ✓

(d) Tollens' reagent ✓ reacts with propanal forming a silver mirror. ✓ [2]

[Total: 8]

2 (a) (i) Solid ✓ [1]

(ii) Alcohol: $CH_2OHCHOHCH_2OH$; ✓ sodium salt: $CH_3(CH_2)_{16}COO^-Na^+$ ✓;
use: soap ✓ [3]

(b) (i)

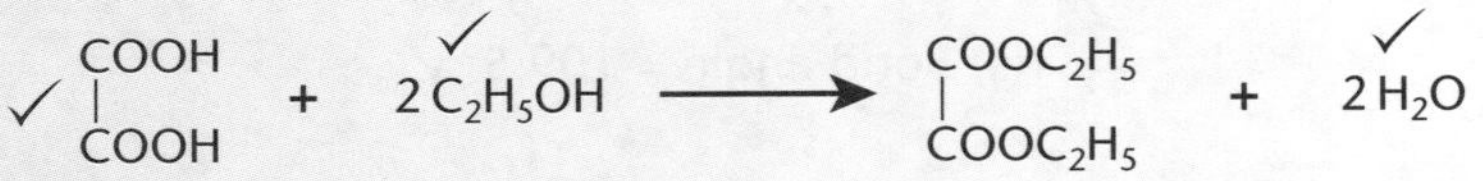

[3]

(ii) Concentrated H_2SO_4 ✓ [1]

(iii)

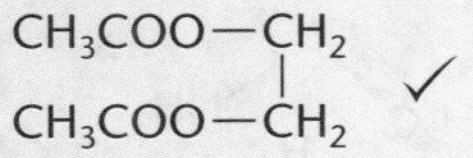

[1]

[Total: 9]

3 (a) (i) NaCN/dilute acid ✓ (ii) $HCl(aq)/H_2O$ ✓ [2]

(b) Chiral or asymmetric carbon atom ✓ [1]

(c) (i) $CH_3COCOOH$ ✓ [1]

(ii) $CH_3COCOOH + 6[H] \longrightarrow CH_3CHOHCH_2OH + H_2O$ ✓✓ [2]

[Total: 6]

6 Aromatics, amines, amino acids and polymers

1 (a) (i) molar ratio C:H = 90.56/12 : 9.44/1 ✓ = 1:1.25 ✓ = 4:5 [2]

(ii) C_8H_{10} ✓ [1]

(iii)

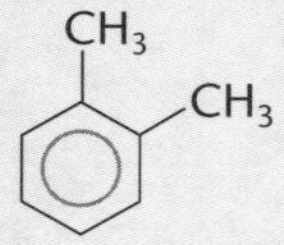

✓

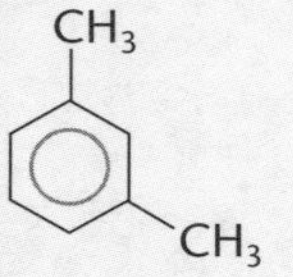

✓

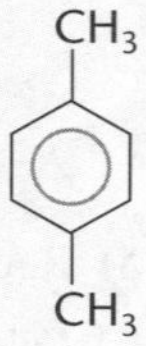

✓

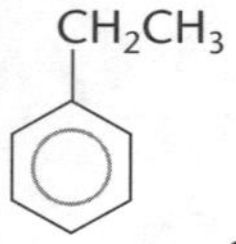

✓

[4]

(b) (i) 1,4-dimethylbenzene ✓ [1]

(ii)

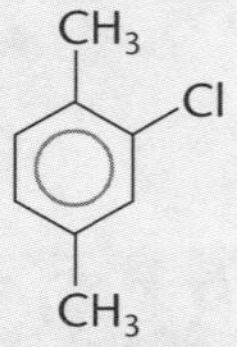

✓

[1]

[Total: 9]

2 (a) A lone pair of electrons on the nitrogen atom ✓ accepts a proton. ✓ [2]

(b) The lone pair has less electron density ✓ due to the negative inductive effect from the aryl ring. ✓ [2]

(c) (i) Nucleophilic substitution ✓ [1]

(ii) $(CH_3CH_2)_2NH$ ✓ $(CH_3CH_2)_3N$ ✓ $(CH_3CH_2)_4N^+Br^-$ ✓ [3]

(d) $CH_3CN + 4[H] \longrightarrow CH_3CH_2NH_2$ ✓
Only one product is formed ✓ [2]

[Total: 10]

3 (a) (i) A chiral carbon atom has **four** different groups attached to it. ✓ [1]

(ii) It provides two possible forms that are non-superimposable mirror images of one another. ✓ [1]

(iii) Bond angle = 109.5° ✓

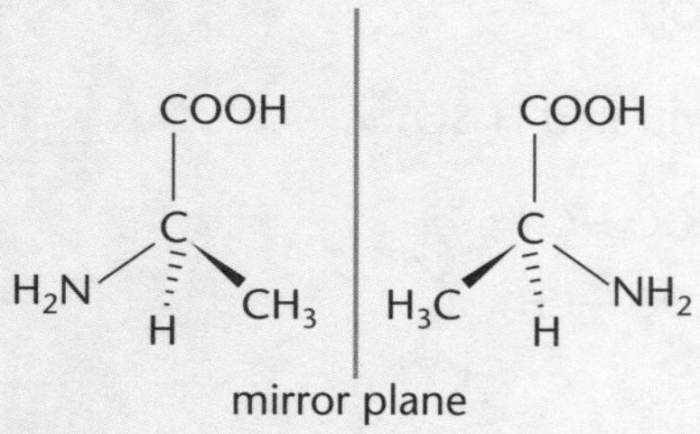

✓✓

[3]

(b)

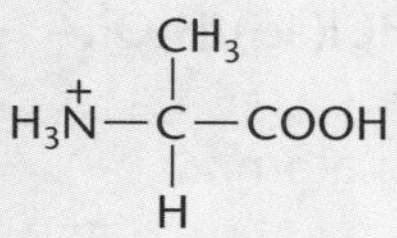

pH = 2.0 ✓

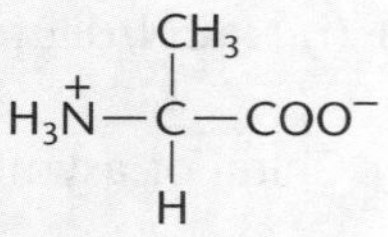

pH = 6.0 ✓

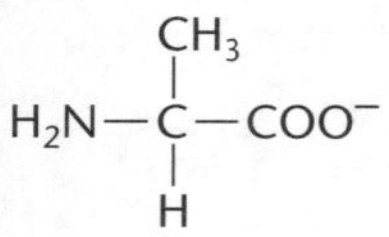

pH = 10.0 ✓

[3]

(c) (i)

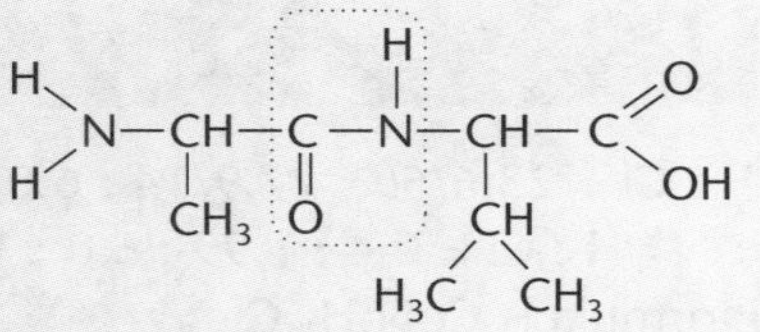

✓ [1]

(ii)

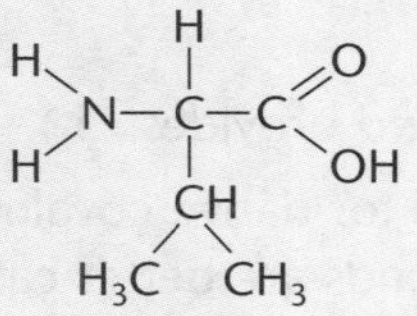

✓ [1]

(iii)

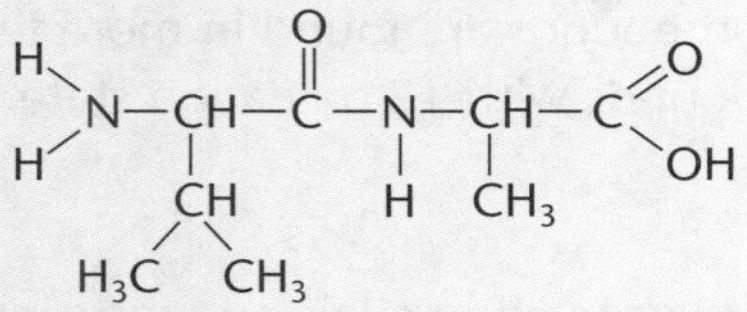

✓ [1]

[Total: 11]

7 Analysis and synthesis

1 (a) **J** 1750 cm^{-1} C=O ✓ ; 1250 cm^{-1} and 1050 cm^{-1} C–O ✓
Evidence suggests an ester ✓
K broad 2500–3300 cm^{-1} O–H (carboxylic acid) ✓
1720 cm^{-1} C=O; 1240 cm^{-1} C–O ✓
Evidence suggests a carboxylic acid ✓ [6]

(b) Molar ratio C:H:O = 54.5/12 : 9.1/1 : 36.4/16 ✓ = 2:4:1 ∴ C_2H_4O✓ [2]

(c) J = $C_4H_8O_2$ ✓ 4 possible structures, e.g. $CH_3CH_2COOCH_3$ ✓ [2]

[Total: 10]

2 (a) Trimethylsilane (TMS) ✓ [1]

(b) 4 ✓ [1]

(c) From left to right, 2:1:6:3 ✓ [1]

(d) C_2H_5 ✓
The CH_2 group at δ 1.53 is split into a quartet ✓ (3 adjacent protons) ✓
The CH_3 group at δ 0.93 is split into a triplet ✓ (2 adjacent protons) ✓ [5]

(e) 6 equivalent protons (2 equivalent CH_3 groups) ✓ [1]

(f) $CH_3CH_2COH(CH_3)_2$ ✓ [1]

[Total: 10]

8 Synoptic assessment

1 (a) Co : N : H : Cl = 23.6/59 : 27.9/14 : 6.0/1 : 42.5/35.5 ✓
Ratio Co : N : H : Cl = 0.4 : 1.99 : 6.0 : 1.2 = 1 : 5 : 15 : 3 ✓
Empirical formula is $CoN_5H_{15}Cl_3$ ✓ [3]

(b) (i) Air is an oxidising agent, ✓ (oxidising Co^{2+} to Co^{3+}) [1]

(ii) NH_3 is ligand ✓ [1]
NH_4^+ is buffer ✓ (and provides Cl^-) [1]

(c) Cobalt uses 3d orbitals for dative covalent bond formation; calcium does not. ✓
Cobalt and its compounds are better catalysts than calcium. ✓
Cobalt compounds coloured, calcium are colourless ✓
Cobalt compounds are found in more than one oxidation state; calcium only has compounds with +2 oxidation state. ✓ [4]

[Total: 10]

2 (a) (i) Long hydrocarbon chain suggests insolubility in water ✓ [1]
OH groups suggest solubility in water ✓ [1]

(ii) 4 geometric isomers ✓
cis:*cis* *cis*:*trans* *trans*:*cis* and *trans*:*trans* ✓ [2]

(iii) with Na:
$H_3C–(CH_2)_2–CH=CH–CH=CH–(CH_2)_8–CH_2O^-Na^+$ ✓
with $H_2SO_4/K_2Cr_2O_7$:
$H_3C–(CH_2)_2–CH=CH–CH=CH–(CH_2)_8–CHO$ ✓
with H_2SO_4/KBr:
$H_3C–(CH_2)_2–CH=CH–CH=CH–(CH_2)_8–CH_2Br$ ✓
with Br_2:

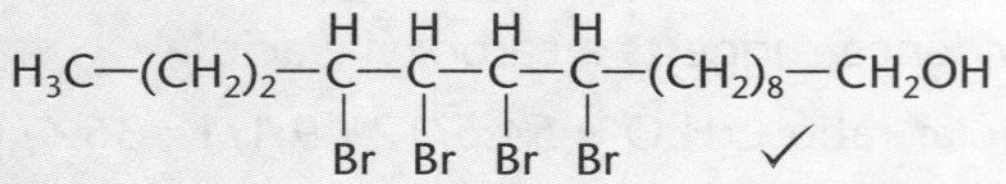

[4]

(b) $CH_3(CH_2)_2COOH$; $(COOH)_2$ and $HOOC(CH_2)_8COOH$ ✓✓ [2]

[Total: 10]

3 Reduction is gain of electrons ✓ decrease in oxidation number ✓
Oxidation is loss of electrons ✓ increase in oxidation number ✓
Inorganic:
$Na + Cl_2 \longrightarrow 2NaCl$ ✓✓
Na oxidised; Cl reduced ✓
$5Fe^{2+}(aq) + MnO_4^-(aq) + 8H^+(aq) \longrightarrow Mn^{2+}(aq) + 4H_2O(l)$ ✓✓
Fe oxidised; Mn reduced ✓
Organic:
Oxidation of alcohol with $H_2SO_4/K_2Cr_2O_7$ ✓
$CH_3CH_2OH + [O] \longrightarrow CH_3CHO + H_2O$ ✓✓
Reduction of nitrobenzene with Sn/HCl ✓
$C_6H_5NO_2 + 6[H] \longrightarrow C_6H_5NH_2 + 2H_2O$ ✓✓
Clear, well-organised, using specialist terms ✓

[Total: 14 max]

4 NaCl has a giant lattice ✓ with strong forces between ions giving a high boiling point. ✓ In solution, the ions are free to move ✓ and conduct electricity ✓
CH_3COOH, CH_3CH_2OH and $AlCl_3$ have simple molecular structures ✓ with weak forces between molecules giving a low boiling point. ✓

In solution, CH_3CH_2OH has mobile uncharged molecules ✓ which cannot conduct electricity. ✓
CH_3COOH partially dissociates in solution giving a small proportion of mobile ions and slight conductivity. ✓
$CH_3COOH \rightleftharpoons CH_3COO^- + H^+$ ✓
$AlCl_3$ conducts by reacting with water:
$AlCl_3 + 3H_2O \longrightarrow Al(OH)_3 + 3H^+ + 3Cl^-$ ✓
Reaction forms mobile ions which can conduct electricity ✓

Solutions of NaCl and CH_3CH_2OH have pH = $-\log(1 \times 10^{-7}) = 7$ ✓
Solution of $AlCl_3$ has pH = $-\log(3 \times 10^{-1}) = 0.5$ ✓
For solution of CH_3COOH,
$K_a = [CH_3COO^-]\,[H^+]/[CH_3COOH]$ ✓
Solution of CH_3COOH has a pH = 2.89 ✓
Clear, well-organised throughout, using specialist terms ✓

[Total: 14 max]

5 (a) $RBr + NaOH \longrightarrow ROH + NaBr$ ✓
$Ag^+(aq) + Br^-(aq) \longrightarrow AgBr(s)$ ✓ [2]

(b) Use a known mass of the haloalkane ✓ with an excess of NaOH(aq) ✓ in a pear-shaped flask ✓ fitted with a condenser ✓ heat ✓ and reflux for 40 minutes. ✓ [5 max]

(c) Acidify with $HNO_3(aq)$ ✓ and add an excess ✓ of $AgNO_3(aq)$. ✓
Filter the precipitate of AgBr ✓ using weighed filter paper, wash with distilled water, ✓ dry the filter paper + AgBr and weigh. ✓
Mass of AgBr = (mass of filter paper + AgBr) – mass of filter paper ✓ [6 max]

(d) Mass of Br in AgBr, b = 80 × mass AgBr/188. ✓
% by mass of Br in haloalkane = b × 100/mass bromoalkane ✓ [2]

[Total: 15]

The Periodic Table

Key: relative atomic mass 1.0; atomic symbol H; atomic number 1; name Hydrogen

1	2											3	4	5	6	7	0
																	4.0 He 2 Helium
6.9 Li 3 Lithium	9.0 Be 4 Beryllium											10.8 B 5 Boron	12.0 C 6 Carbon	14.0 N 7 Nitrogen	16.0 O 8 Oxygen	19.0 F 9 Fluorine	20.2 Ne 10 Neon
23.0 Na 11 Sodium	24.3 Mg 12 Magnesium											27.0 Al 13 Aluminium	28.1 Si 14 Silicon	31.0 P 15 Phosphorus	32.1 S 16 Sulphur	35.5 Cl 17 Chlorine	39.9 Ar 18 Argon
39.1 K 19 Potassium	40.1 Ca 20 Calcium	45.0 Sc 21 Scandium	47.9 Ti 22 Titanium	50.9 V 23 Vanadium	52.0 Cr 24 Chromium	54.9 Mn 25 Manganese	55.8 Fe 26 Iron	58.9 Co 27 Cobalt	58.7 Ni 28 Nickel	63.5 Cu 29 Copper	65.4 Zn 30 Zinc	69.7 Ga 31 Gallium	72.6 Ge 32 Germanium	74.9 As 33 Arsenic	79.0 Se 34 Selenium	79.9 Br 35 Bromine	83.8 Kr 36 Krypton
85.5 Rb 37 Rubidium	87.6 Sr 38 Strontium	88.9 Y 39 Yttrium	91.2 Zr 40 Zirconium	92.9 Nb 41 Niobium	95.9 Mo 42 Molybdenum	– Tc 43 Technetium	101 Ru 44 Ruthenium	103 Rh 45 Rhodium	106 Pd 46 Palladium	108 Ag 47 Silver	112 Cd 48 Cadmium	115 In 49 Indium	119 Sn 50 Tin	122 Sb 51 Antimony	128 Te 52 Tellurium	127 I 53 Iodine	131 Xe 54 Xenon
133 Cs 55 Caesium	137 Ba 56 Barium	139 La 57 Lanthanum	178 Hf 72 Hafnium	181 Ta 73 Tantalum	184 W 74 Tungsten	186 Re 75 Rhenium	190 Os 76 Osmium	192 Ir 77 Iridium	195 Pt 78 Platinum	197 Au 79 Gold	201 Hg 80 Mercury	204 Tl 81 Thallium	207 Pb 82 Lead	209 Bi 83 Bismuth	– Po 84 Polonium	– At 85 Astatine	– Rn 86 Radon
– Fr 87 Fracium	– Ra 88 Radium	– Ac 89 Actinium	– Rf 104 Rutherfordium	– Db 105 Dubnium	– Sg 106 Seaborgium	– Bh 107 Bohrium	– Hs 108 Hassium	– Mt 109 Meitnerium	– Unn 110 Ununnilium	– Uuu 111 Unununium	– Uub 112 Ununbium		– Uuq 114 Ununquadium		– Uuh 116 Ununhexium		– Uuo 118 Ununoctium

lanthanides	140 Ce 58 Cerium	141 Pr 59 Praseodymium	144 Nd 60 Neodymium	– Pm 61 Promethium	150 Sm 62 Samarium	152 Eu 63 Europium	157 Gd 64 Gadolinium	159 Tb 65 Terbium	163 Dy 66 Dysprosium	165 Ho 67 Holmium	167 Er 68 Erbium	169 Tm 69 Thulium	173 Yb 70 Ytterbium	175 Lu 71 Lutetium
actinides	– Th 90 Thorium	– Pa 91 Protactinium	– U 92 Uranium	– Np 93 Neptunium	– Pu 94 Plutonium	– Am 95 Americium	– Cm 96 Curium	– Bk 97 Berkelium	– Cf 98 Californium	– Es 99 Einsteinium	– Fm 100 Fermium	– Md 101 Mendelevium	– No 102 Nobelium	– Lr 103 Lawrencium

Index